Advance Praise for *Iconic Advantage*®

"This book explains why some brands are built to last and others seem doomed to perish. It's a framework that every marketer can put into play right away."

—Adam Grant, *New York Times* bestselling author of *Give and Take*, *Originals*, and *Option B* with Sheryl Sandberg

"Soon has an uncanny ability to take mysteries and turn them into heuristics. He's done it on innovation and design, and now with *Iconic Advantage*."

—Roger Martin, bestselling author of *Playing to Win* and Former Dean of the Rotman School of Business

"Great brands are truly iconic...but creating one is as much science as it is art. This book teaches you the key ingredients so you can create your own Iconic Advantage."

—Claudia Kotchka, Former VP of Design Innovation & Strategy, Procter & Gamble

"Having built a $2 billion pipeline of innovation for brands like Van's, Timberland, The North Face, and Wrangler, Soon understands how to create iconic products...if you aspire for your brand or products to be iconic, read this book"

—Chip Heath, *NYT* bestselling author of *Switch* and *Made to Stick*, Professor, Stanford Graduate School of Business

"Icons are critical strategic assets for brands and companies. This book offers deep insights on how to establish timeless distinction and relevance, and builds a compelling case for all companies, from Fortune 500 to venture-backed startups, to focus their resources on building greater iconicity."

—Iris Yen, Global VP of Strategy, Nike

"*Iconic Advantage* is even more important for startups than it is for established brands and large companies."

—Wen Hsieh, General Partner, Kleiner Perkins Caufield Beyers

ICONIC®
ADVANTAGE

ICONIC®
ADVANTAGE

Don't chase the new, innovate the old

Soon Yu
with Dave Birss

A SAVIO REPUBLIC BOOK
An Imprint of Post Hill Press

Iconic Advantage®
© 2018 by Soon Yu
All Rights Reserved

ISBN: 978-1-68261-540-9
ISBN (eBook): 978-1-68261-541-6

Cover Design by Dave Birss and Fiona Zwieb
Interior Design and Composition by Greg Johnson/Textbook Perfect

posthillpress.com
New York • Nashville
Published in the United States of America

To Christine and Brenden,
for making me laugh when I got too serious
and for inspiring me when I had doubts.

—SOON

For Valerie, Iona, and Simone
for their incredible tolerance
of a temperamental writer.

—DAVE

Contents

CHAPTER 1

The Importance of Being Iconic

In 1989, Dov Charney had an idea in his university dorm room. It wasn't a big idea. It was simply exporting American-made T-shirts to Canada. He called his company American Apparel. Very soon he found himself making enough money to drop out of university and pursue the business.

Within 10 years he'd moved to LA, opened up a manufacturing plant, and shaken up the garment industry by paying his workers twice the going rate. In a few years, American Apparel grew to become the largest T-shirt manufacturer in the US.[1]

Seeing the popularity of his products, he expanded the company from wholesale to retail, eventually opening over 250 outlets around the world.[2]

He grabbed the world's attention with a controversial advertising strategy. American Apparel's ads looked more like spreads from *Playboy* magazine than traditional fashion adverts. It was often surprising how few products the models were wearing. Sales continued to rise.

However, the company's racy ads began to generate more attention than the products they were advertising. If you mentioned the brand in conversation, people would talk about the sexually explicit

imagery rather than the quality of the clothing and its ethical manufacturing.

In an interview with American Public Media radio, Dov Charney even admitted, "My biggest weakness is me. I mean, lock me up already. It's obvious. Put me in a cage, I'll be fine. I'm my own worst enemy. But what can you do? I was born strange."[3]

His "strangeness" began to be a significant liability for the business. It generated poisonous rumors and sexual misconduct lawsuits. The brand no longer represented what the market wanted. It had completely lost its relevance.

Imagery that was initially seen as sexually adventurous was now seen as sleazy. The higher price of the garments didn't stack up against the cheaper high-street retailers. And their staunchly hipster aesthetic wasn't attracting new customers.

In 2014 there was a mutiny. The board of directors ousted Dov Charney from his own company. They hoped that returning to the basics would help the company rebuild itself. But it was too little, too late. In October 2015, the company filed for bankruptcy.[4]

At one point in time, the brand appeared to have everything going for it. But it didn't last. And the business hit the dirt in a rather dramatic way.

So what went wrong?

Like many rapidly growing brands, American Apparel relied on market momentum and noisy advertising. An old adage is that advertising is the tax you pay for having an unremarkable product. And no amount of advertising can help you when you're not sufficiently differentiated from a new wave of cut-price fashion retailers.

As far as we can see, American Apparel failed to refresh its product line to remain relevant to its loyal customers. As the fashion website *Racked* put it: "American Apparel are trying to sell an aesthetic that is no longer fresh and exciting."[5]

In short, the company neglected to enhance its **Noticing Power** to remain distinctive and bolster its **Staying Power** to remain relevant. Without doing that, it was unable to create any long-term **Iconic Advantage®**.

American Apparel ended up being a fad rather than an iconic brand. Too many businesses fall into that trap. It will be interesting to see if the brand's new owners can get the company's mojo back.

If you had to list the icons of swinging '60s London, you'd probably include the Mini Cooper somewhere near the top. It was a car that really seemed to capture the spirit of the time. It didn't look like anything else on the road. It was quirky, cool, and refreshing. And it's no surprise that it became such a success.

However, as the swinging '60s made way for the sexy '70s, the electronic '80s, and the naughty '90s, the Mini ran out of road.

The brand failed to capitalize on the amazing iconic power of the product that was right under its nose. Over the years, it hadn't really invested in **Staying Power** to keep the brand relevant and meaningful.

The classic Mini

By the end of the century, the last car it ever made, the 5,387,862nd, was almost identical to the very first one it made.[6]

Fortunately, this wasn't the end of the story. BMW was standing by to turn this 20th-century icon into a 21st-century one.

BMW bought the iconic brand, having seen the opportunity to revive it for the modern market. Working with its design consultancy team, Designworks, BMW set to work identifying and defining everything that had made it so iconic. And it did so with a quirkiness that perfectly matched the brand.

It created a "human body archetype" that merged the cuteness of a child, the muscular shoulders of a man, and the soft forms of a woman. This added an emotional, human element to the 3D design of the car.[7]

It identified a handful of iconic elements—the oval headlights, the hexagonal radiator grille, and the elliptical door handles—and it highlighted them further by adding chrome surrounds.[8] This helped to draw attention to the distinctive elements that had given the original Mini its **Noticing Power**.

The BMW-designed Mini

With this iconic groundwork done, BMW set about designing an updated version of the car that retained the classic character but was updated to fit the higher expectations of the modern driver. This more relevant design laid the foundation for the car's **Staying Power.**

For the second time in its history, the Mini became a hit.

BMW went on to add the final piece of the puzzle by building the Mini's **Scaling Power** to boost its recognition. It did this by expanding the line of products from a single model to a five-door wagon, a convertible, and several other variants. It took the **Iconic Brand Language**™ it had developed and used it to inspire a broad range of Mini merchandise, from watches to luggage to sportswear.[9] And finally, it used advertising and promotional channels to give the Mini maximum visibility for its audience.

This is an absolute master class in iconic design.

This book reveals how the world's biggest brands are using the above approach. And it shows you how you can apply that thinking to give your own organization Iconic Advantage.

A lot of established companies are now operating in unfamiliar territory. The world of business has always evolved and developed, but today's accelerating pace of change has already surpassed the top speed of most companies. And it's sending them into a spin as they do their best to catch up. In many industries, the market is changing faster than organizations can. More energy is being spent on plugging leaks than plotting a course.

Entire markets are being pushed towards commoditization as consumers are demanding things cheaper and faster. The phenomenon of "fast fashion" has transformed the high street into a place that delivers quantity rather than quality. It's all about low margins and speed to market. And that offers little buffer against market fluctuations.

Another side effect of commoditization is reduced loyalty. When people make decisions based on the number of items they can get for the cash in their pocket, they're more inclined to walk across the road and give their money to a competitor. Retailers are currently trying to fix that by investing in social media in the hope that building relationships with consumers leads to loyalty and increased sales. The jury is out on whether that's effective.

And traditional marketing isn't the brand-building solution it once was. Recent years have seen most forms of advertising become less and less effective. At the same time, media channels have exploded. The only way consumers can deal with the increasing deluge of marketing messages is to filter them out. And that's making it harder than ever for brands to communicate with their audience.

Many corporations are still struggling on with their 5-year, 10-year, or 20-year plans, regardless of the fact that their circumstances have changed since they wrote them. In the last ten years, we've seen the rise of

> *Companies find themselves reacting to circumstances rather than creating them. They lose control, they lose vision, and—ultimately— they lose profits.*

social media, the growth of cloud services, and an explosion of video content. No one can predict what effect virtual reality, cognitive computing, and self-driving vehicles will have in the coming years. And what unexpected technology is waiting just around the corner.

So companies find themselves reacting to circumstances rather than creating them. They lose control, they lose vision, and—ultimately—they lose profits.

However, there are always exceptions to the rule. We researched dozens of companies that seem to transcend the issues that their counterparts found insurmountable. These exceptions are less affected by the fickleness of the market and the rapid advancement of technology. They continue to build loyalty and expand into new markets. And they seem to grow from strength to strength, year after year.

What's their secret? And how do other companies emulate this secret?

That's what this book is all about. It's about a strategy that these companies use to create lasting differentiation and build deeper relationships with customers. It's about unlocking the value that's hidden away in existing products and services. We call it Iconic Advantage. And we want to show you how you can use it to transform any business.

What Is Iconic Advantage®?

Iconic Advantage is a strategy that focuses on building time-less distinction and relevance so businesses can rise above their competition and build stronger emotional connections with their audience. It's a strategy that applies across every area of an organization. And it's a strategy that helps them stay focused on what matters year after year.

This strategy increases the three key qualities that make products and services iconic:

▶ **Distinction:** Iconic products are known for something distinctive and memorable. This allows them to stand out.

▶ **Relevance:** Iconic products are not just different for difference's sake. Their distinction is highly relevant and meaningful to their audiences. This allows them to stick around.

▶ **Recognition:** Iconic products gain universal recognition for their distinctive relevance. They become the stan-dard-bearer in their category, segment, or niche for this distinctive relevance.

The iconic part is important. Like religious icons and cultural icons, we're talking about properties that have meaning, rele-vance, and an emotional connection with an audience. These are properties that have become the standard-bearer for their category, niche, segment, or movement. We can learn as much from Andy Warhol and NSYNC as we can from Apple on this subject. It's this stronger human connection that helps to create a real, sustainable business advantage in the market. It gener-ates true loyalty and can result in sales volumes that drive costs down and multiply profit margins.

This isn't anything particularly new. Many of the world's most successful companies have been benefiting from this strategy for years. And throughout the book, we'll be showing you examples of how they've done that.

> *Like religious icons and cultural icons, we're talking about properties that have meaning, relevance, and an emotional connection with an audience.*

We'll be sharing lessons from household names like Nike, Disney, Amazon, and In-N-Out Burger. You'll see how these companies defined what it was that differentiated them from their competition. And you'll see how they created a process to protect this essence and use it to expand their business over time.

These companies didn't stumble across their Iconic Advantage by accident. It was a conscious strategic decision. We'll show you how any business can take advantage of it, whether you are a Fortune 500 or a venture capital-backed startup. And regardless of whether you're selling products or services.

Best of all, a lot of businesses already have the ingredients they need right under their noses. They've got products with existing audiences, a distribution network, and retail outlets. Or they've got services with an existing user base and presence in the market. Iconic Advantage helps them unlock additional value that leads to increased profit and sustained growth.

You can also use the principles to guide you in creating new iconic properties that have a much higher chance of success.

It's important to note that Iconic Advantage isn't an instant fix. It's a cumulative effect that gains power over time—much in the same way as Jimi Hendrix wasn't a guitar legend when he first picked up a guitar. It took time and effort to get there. This book is all about putting you on the right path and giving you the best chance of success.

What Iconic Advantage Isn't

Iconic Advantage isn't just branding or design. These play an important role in bringing Iconic Advantage to life, but this is a strategy that runs deeper than aesthetics.

Likewise, it's not an advertising or marketing strategy. Of course, marketing plays a role in communicating with the audience, but in many ways, Iconic Advantage is about baking the desirability into the product or service rather than expecting a bunch of posters and TV ads to convince your audience to part with their cash.

Iconic Advantage isn't research and development (R&D). Again, it embraces elements of R&D, but it's more about focusing your efforts on creating benefits that resonate with the audience. It puts the audience before the product and emotional benefits before rational features.

Rather than being a strategy that's embraced by just one part of an organization, an Iconic Advantage strategy involves every area of the company. It needs to influence marketing, branding, design, finance, mergers and acquisitions, R&D, merchandising, sales, innovation, and organizational structure. Each of these disciplines needs to consider how it will implement the strategy. It's not something that happens by chance. It's a conscious decision that involves the entire organization, from the management team to the interns.

A Simple Approach to Strategy

Don't let the word "strategy" put you off. Strategy has become overcomplicated in recent years. It's seen as the preserve of geniuses with a superhuman ability to spot opportunities no other mortal can see. But it's a lot simpler than most people believe.

Strategy is nothing more than principles that guide a series of decisions. Roger L. Martin and A. G. Lafley describe a clear framework for effective strategy in their book *Playing to Win*. It follows a series of choices that can lead you from high-level aspirations to implementation within your business.[10] Here's how Iconic Advantage fits into that framework.

Iconic Advantage Strategy

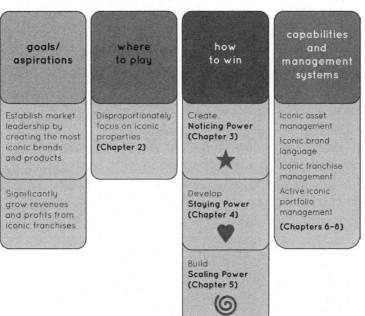

goals/ aspirations	where to play	how to win	capabilities and management systems
Establish market leadership by creating the most iconic brands and products	Disproportionately focus on iconic properties **(Chapter 2)**	Create **Noticing Power (Chapter 3)** ★	Iconic asset management Iconic brand language Iconic franchise management
Significantly grow revenues and profits from iconic franchises		Develop **Staying Power (Chapter 4)** ♥	Active iconic portfolio management **(Chapters 6–8)**
		Build **Scaling Power (Chapter 5)** 🌀	

Goals and Aspirations

Iconic franchises tend to be significantly more profitable than standard ones. They're also far more resilient in the face of market fluctuations. But you can't create an iconic property overnight. Iconic Advantage is a compound effect that you earn after a period of deliberate and sustained commitment. The effect is stronger over time. So the only way to start is to choose to become iconic. An Iconic Advantage strategy is for companies that want to establish market leadership by becoming the standard-bearer in their market.

Where to Play

This is all about focusing your efforts in the right places. It means instead of just chasing shiny new objects, you focus on innovating the old—where your existing strengths lie. Iconic Advantage involves overindexing your efforts on your iconic properties. On the flip side, it also means that you reduce your focus on properties with low iconic potential, or even divest them so that you can concentrate your efforts on the properties with the best chance of success. You need to put your energy and resources in the right place.

This sentiment, which is covered in Chapter 2, can be summed up by the advice that Steve Jobs gave to Mark Parker, the CEO of Nike:

"Nike makes some of the best products in the world. Products that you lust after. But you also make a lot of crap. Just get rid of the crappy stuff and focus on the good stuff."[11]

How to Win

To create something iconic, you need to stand out from the competition. And this is the soul of your Iconic Advantage. Our goal is to supercharge the three qualities that make a product iconic: distinction, relevance, and universal recognition. We're going to be picking each of these apart and explaining how to tackle them in more detail. This will help you create a property that's different from your competitors', that connects on a deeper level with your audience, and that opens up growth opportunities for your business.

The three steps are as follows:

1. **Create Noticing Power**

    It's impossible to stand out by being the same. Yet, most markets end up filled with generic products with a similar list of features. You need to make your offering **distinctive**. You need to stand out. The best iconic products look markedly different from the other products on the shelf. Their difference is appealing and meaningful, rather than just being different for the sake of it. Once you've stood out and gotten people's attention, you've increased the chances that your audience will also notice your advertising, promotions, merchandising, partnerships, and other activities. You'll learn how to do this in Chapter 3.

2. **Develop Staying Power**

    You can't build an iconic brand without developing a deeper connection with your audience. You need to make it **relevant** so it sticks around. The best kind of iconic brand has a story to it. This can be the heritage of the brand. Or standing for something radically different from the rest of the market. Or the breakthrough science that delivers a meaningful benefit. This is what gives the brand a deeper connection with the audience that you could never achieve with mere aesthetics. This is covered in Chapter 4.

3. **Build Scaling Power**

   Once you've got the first two principles in place, you need to give your iconic property as much presence as possible to become **universally recognized**. To do this, you start with a strongly defined Iconic Brand Language. This becomes the bible that guides all your decisions as you expand your iconic property. You can use this guide to expand your franchise into new segments and channels. And then you can drive awareness further through your marketing and promotion. Chapter 5 explains this in more detail.

These three steps create distinction, relevance, and recognition, which form the core of an Iconic Advantage strategy.

Capabilities and Management Systems Required

As you now know, an Iconic Advantage strategy embraces every area of the business. So the capabilities that are required to make it succeed are pretty wide-ranging. It requires empowering leaders with the responsibilities, resources, and tools to deliver the strategy. Without accountability, clear goals, and deliverables, it's hard for people to know if they're succeeding.

In the same way that you can't expect the same recipe to produce a different meal, you need to develop different processes and structures to create Iconic Advantage. The capabilities and management systems required are covered in Chapters 6 to 8.

More than Just a Theory

Iconic Advantage isn't just academic theory. It's a practical strategy that businesses in every sector can use to generate long-term sustainable profit and growth.

Much of the thinking comes from the experience of the authors, who bring their different perspectives to the topic.

Soon has over twenty years' experience in strategy, innovation, design, and branding, working for household names as well as starting up his own businesses. Throughout his career he's always been fascinated with what makes some companies more successful than others at creating a meaningful and lasting impact. That led to his leading a study of over 50 companies to unearth the reasons why some brands don't just stand the test of time but, also reach iconic status. It's the findings of that study—and the thinking it prompted—that led to this book.

Dave came on board as a collaborator and case writer to add meaningful stories and research to illustrate the thinking. He also brought twenty years of advertising experience, from every area of the marketing mix, along with scientific geekiness and a healthy dose of Scottish contrariness.

As part of the writing journey, we've worked together to apply Iconic Advantage thinking to the needs of actual clients. Because nothing tests a theory like real-world application.

The Iconic Advantage approach comes from a distillation of the strategies of some of the world's most successful organizations. Brands like Nike, BMW, In-N-Out Burger, Disney, and even the Catholic Church have benefited from this approach. And in this book, you'll find stories of how they've used an Iconic Advantage strategy to rise above their market and continue to grow year after year.

This book doesn't just explain what Iconic Advantage is—it explains how you can use it to transform your own organization. You'll find frameworks and methodologies—and even some downloadable worksheets that will help to direct your thinking. This is about doing, not just talking.

If you're looking for more reasons to embrace Iconic Advantage, you'll find plenty of them in the next chapter. And let's begin by taking a look at the business case for Iconic Advantage.

CHAPTER 2

The Case for
Iconic Advantage®

In 2015, China's ride-sharing market was a pretty hostile and competitive environment. It was dominated by DiDi, the Beijing-based market leader, and was seeing rapid growth from Uber, the Silicon Valley darling. The market had become so competitive that over half of the trial rides were being offered free. Most businesses wouldn't have seen an opportunity, especially entering the market as the fifth competitor.

But upstart Ucar wasn't like most businesses. It's run by Charles Zhengyao Lu, a business maverick with a reputation for building iconic automotive empires. He knew the company needed to offer something different if it was to have a chance of succeeding. So Ucar collaborated with Continuum, a global innovation design agency, to dig deeper into the market and identify unmet customer needs.

The Continuum team spent time with atypical riders to try to tease out gaps and opportunities in the current service. This process identified growing concerns about reliability and passenger safety. The rapid growth in the market meant that the big companies were having difficulty controlling the quality of drivers and vehicles. The same app could provide a shiny black limo with a suited driver one day and a dirty pickup truck with a disheveled driver wearing flip-flops the next. If Ucar could remove this pain point, it would have a distinctive benefit for the growing Chinese market.

It also had a strength it wanted to leverage. Ucar was built on the foundation of a successful car rental business, which meant it already had a fleet of vehicles and an existing team to keep them in the best condition. It wanted to innovate on this existing infrastructure for its new venture. So it used these insights and resources to develop a service with some outstanding signature elements that communicate the key points of difference.

Ucar offers consistently high-quality cars that are clean and black, with Wi-Fi, chargers, and water for the passengers. The drivers arrive smartly outfitted with ties, coats, and white gloves. They have a specific way of greeting their passengers, opening the door for them like a high-class chauffeur. There's nothing standard about this service.

The company also created signatures that helped embody safety and reliability. The drivers are fully salaried and constantly monitored to make sure they keep to speed limits and deliver reliable, safe experiences. And being on the payroll means they can concentrate on the customer experience rather than maximizing the number of rides they can squeeze into an 18-hour shift.[1]

Ucar is completely focused on reliability, safety, and trust. This has allowed it to create meaningful **Noticing Power** and lasting **Staying Power**. It has chosen to deliver an outstanding service experience to a smaller market. And that allows it to make more money for fewer rides.

It has gone even further for the very niche market of mothers-to-be. If a pregnant woman orders a Ucar, she'll get one of its highest-quality cars with a driver who has at least three years' experience plus a knowledge of basic medical treatments for pregnant women. On top of that, the cars are fitted with additional air purifiers, are kept at a constant 22 degrees Celsius (about 71 degrees Fahrenheit), and don't go faster than 60 kilometers per hour (about 37 miles per hour). The drivers even play baby music for the little passenger-to-be.

Ucar and Continuum started with the intention of creating an iconic brand. And they did it brilliantly by creating **Noticing Power** and **Staying Power** from the get-go. Ucar has successfully become the standard-bearer for safety and reliability in China's ride-sharing market.

This has paid off massively. In the last two years, it has doubled the number of cities it serves, quadrupled the number of drivers, and increased its average fare per ride by 50 percent. On top of that, it consistently scores the highest in customer satisfaction and retention versus any of its competitors.

Few companies can teach you how to pick up customers better than Ucar can.

If you need any more convincing about the power of Iconic Advantage®, read on. We'll be looking at it from financial, organizational, theoretical, psychological, and sociological points of view. We'll look at the benefits you can hope to achieve with an Iconic Advantage strategy. And we'll explain what it is that can affect the success of Iconic Advantage in your organization.

As we've outlined in Chapter 1, the strength of Iconic Advantage is in creating stronger connections with your audience. So a good place to start is by examining how humans respond to brands and then see how Iconic Advantage can help you strengthen that bond.

The Ingredients of Brand Equity

First of all, let's define what we mean by "brand equity."

Contrary to what most companies believe, they don't actually own their own brand. It's not the sum of their logos, colors, fonts, graphical assets, mission statement, tagline, and audio sting. Instead, each brand is something that lives in the minds of its audience. And each individual has a different interpretation of the brand based on his or her experiences and what he or she has heard other people say.

For example, Milton Glazer's famous "I ♥ NY" logo is a visual element that represents New York City.[2] But people's opinions of New York are entirely dependent on their experiences. Someone who was proposed to at the top of the Empire State Building during a romantic weekend visit will have a very different idea of New York than someone who couldn't wait to escape from a tough inner-city neighborhood. The city's government had an influence on both experiences, but it has only so much control over what people think of the city. The idea of what New York is like lives only in the heads of the individuals. And when they tell other people their story of the city, it affects their opinion too.

Your brand is the sum of a person's experiences and beliefs about your company, product, or service.

The "equity" part is anything that's valuable to the brand.

So when we talk about brand equity, we're talking about the common experiences, beliefs, assets, and stories that affect the brand's value in either a good or bad way. These experiences are driven by five factors:

> *Your brand is the sum of a person's experiences and beliefs about your company, product, or service.*

▶ **Uniqueness**

Iconic brands have elements that differentiate them from the competition. These iconic elements help them to rise above commoditized markets and earn a higher share of attention.

▶ **Familiarity**

Your audience knows what your brand is about. And the more familiar they are with it, the more they trust it. Because of iconic brands' recognizability, they have earned an unusually high level of trust.

▶ **Relevance**

If you're going to connect with your audience, your brand needs to relate to its needs and beliefs. Iconic brands go beyond that and become entities that the audience defines itself by.

▶ **Quality**

The brand needs to deliver a consistent level of quality. People need to know what to expect when they use the product. That doesn't mean it has to be top quality. Both BIC pens and Montblanc pens are iconic even although they are at opposite ends of the scale. However, the higher perceived quality of Montblanc pens allows them to sell for hundreds of times more money, while the mass-produced quality of BIC pens allows the company to sell thousands of times more product.

▶ **Popularity**
Iconic brands gain a following of passionate people. These individuals are very often opinion leaders who influence a much bigger group of people. The more people who buy the brand within your community, the more desirable it becomes. Sales lead to more sales, and the brand takes on a meaning far beyond its functional benefits.

Together these factors form a brand equity index that you can use to measure the strength of your brand.[3] In Chapter 3, we'll discuss the importance of **uniqueness** when you're creating Noticing Power. In Chapter 4, we'll outline the best practices required to build familiarity, relevance, and value for greater Staying Power. And lastly, in Chapter 5 we'll explain how to scale your iconic popularity with your audience.

The power of Iconic Advantage is that it encourages organizations to focus on these brand dimensions. By their very definition, iconic properties resonate with an audience and develop a stronger bond with them because they become the standard-bearer for a category feature or benefit. Iconic Advantage is a strategic capability that helps organizations have a better chance of achieving this. And that makes it the most powerful form of branding—a form of branding that extends way beyond the confines of a marketing department to affect the entire organization.

The Different Levels of Branding

Because branding is so poorly misunderstood by the majority of businesspeople, many businesses don't get much value out of it. They limit themselves to the most basic form of visual identity and miss out on the massive opportunities in creating a brand that connects with their audience, generates sales, and protects them against market fluctuations.

Iconic branding goes beyond what your audience sees, thinks, and feels. It reaches a deeper part of what people believe about themselves. The product becomes something the audience members define themselves by.

The power of the brand connection increases as it goes deeper. The different levels of branding go like this:

Iconic branding goes beyond what your audience sees, thinks, and feels. It reaches a deeper part of what people believe about themselves.

Hierarchy of Branding

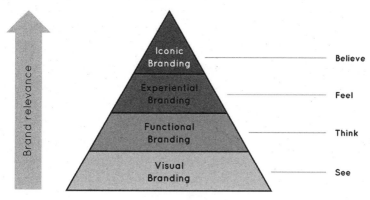

Brand relevance

Iconic Branding	Believe
Experiential Branding	Feel
Functional Branding	Think
Visual Branding	See

▶ **Level 1: Visual branding**

Most brands have some kind of corporate identity guide. It shows the different variants of the logo and outlines the rules of how each one should be used. It lays down the color palette, the fonts, the image style, and any other visual elements. It's all about control. Because the world's top brands tend to have a visual consistency, the hope is that acting like them will open doors to success.

▶ **Level 2: Functional branding**

This is about moving beyond recognition to understanding. You want people to know what it is you do and how that applies to them. Ideally, this starts with your product design. If it communicates what makes your product special at a glance, it makes this stage so much easier. However, that's not always possible. That's why many brands use advertising and other communication channels to let people know why they should buy it.

▶ **Level 3: Experiential branding**

More powerful brands move beyond information to emotion. To reach this level, a brand needs to resonate with the sentiments of its audience. It also needs to understand that actions are more important than words. You can't do this cynically. An audience will see right through that. Instead, it should come from the values that run through the whole organization. Experiences aren't just big PR stunts and high-profile sponsorship deals. The most important experiences are researching, purchasing, unpacking, and using the product or service. And—vitally—anything that involves interaction with your staff.

▶ **Level 4: Iconic branding**

This is the ultimate form of branding, and very few brands operate at this level. This is when the brand goes beyond a superficial connection to its audience and taps into the aspirations and beliefs of who its audience wants to be. These brands transform into icons that the audience uses to define itself. Brands such as Nike, Apple, Porsche, and Harley-Davidson are at this level. And they didn't reach it by accident. There are some very deliberate things a brand needs to do to become an iconic brand.

Iconic Differentiation

Throughout human history, icons have played a central role in society. They were central to the earliest mythology, culture, and religion. And they've never left us since. Iconic visual elements became shorthand for a collection of aspirations, beliefs, and stories. We ended up with symbols, relics, and myths that have played a significant role in people's lives. It may seem archaic to most people, but the human need for icons hasn't diminished. If anything, it has accelerated alongside our media channels.

This is because icons become shorthand for larger concepts and stories. If I asked you to explain what a crucifix represents, you'd struggle to do it justice in a sentence. And some theologians could write entire books on it. But the symbol represents all of that—all of the information, all of the meaning, and all of the emotion. In a world where we are bombarded with increasing volumes of information from constantly multiplying media channels, these potent symbols become even more valuable.

So icons continue to play a vital role for people all over the world, even though most of them are now secular and owned by corporations.

Some brands have such outstanding success at standing for a category benefit that their name becomes synonymous with it. Brands such as Xerox, Google, and Hoover became verbs. Other brands, like JELL-O, Clorox, and Q-tips, become the generic name for their category. That doesn't happen for all iconic brands. But it's a pretty good sign that a product has become one.

On its most basic level, iconic brands become a shortcut for decision-making. In most categories, consumers face a dizzying number of products, all with slightly different benefits. It's a complex job to weigh each option based on rational facts alone. So it's no surprise that most decisions are based on instinct and habit.[4] With their deeper human connection, iconic brands become an easy decision in a crowded marketplace.

That's not a great surprise when you understand how the brain processes images. Part of our visual processing is handled by the amygdalae, a pair of almond-shaped structures deep in the temporal lobes of the brain. These are involved in memory and emotion, providing emotional reactions to visual stimuli. What we see and what we feel are intrinsically linked.[5]

On another level, humans seem to have a psychological need for icons. As Abraham Maslow described in his hierarchy of needs, once people's desires for physiological basics, safety, belonging, and esteem have been met, they aim for self-actualization. They want to be the best version of themselves. And iconic brands work as catalysts to help people reach this ultimate level of being.[6]

Icons have a magical talisman quality that transforms people. They actually put people into a different state just by their using them. Think of how you feel signing a document with a BIC pen compared to signing it with a Montblanc. How about being chauffeured in a high-end Honda compared to being chauffeured in a Rolls-Royce? The functional difference between each option is marginal, but the experiences are incomparable.

In a world where we are bombarded with increasing volumes of information from constantly multiplying media channels, these potent symbols become even more valuable.

Icons also give people a sense of belonging to a community. They feel a kinship with other people who buy the same brand. Mac users feel part of a creative community. Jimmy Choo shoe wearers feel part of a fashion elite. Harley-Davidson riders feel part of a tribe of road warriors. The feeling of belonging is an important human need, and iconic brands fulfill that in spades. They make loyal customers feel connected to a wider community of discerning individuals.

To see the incredible power of iconic brands, it's worth taking a look at what is possibly the oldest brand in history: the Catholic Church.

Its main symbol is so simple and recognizable that children can draw it before they can even write. Yet it's powerful enough to move people to tears (and even repel vampires). You can recognize its uniforms, its architecture, its music, its language, and possibly even its aromas. You can name its principal characters and tell some of its stories. And millions of people use it as a way of defining who they are. Some even go so far as to give their entire life to the brand, dedicating themselves to spreading God's love and helping those in need.

None of this was accidental. It's all defined and protected by the head office in Rome. It's as well-managed as the most successful corporate brands. So from here on in, we'll refer to it as Catholicism™.

For the majority of people, brand Catholicism™ makes them feel something. For many, that is a devoted sense of spiritual love and purpose. So what are the secrets of such massive and long-lasting brand success?

Catholicism™ has some clear visual iconic language. The crucifix as a logo is immediately recognizable. The general shape of a long vertical bar with a shorter horizontal bar is immovable—but beyond that, there's a lot of flexibility. It can be any color you want. It can be ornate or plain. It can have Christ on it or not. There are then even more flexible supporting elements in the uniforms, architecture, iconography, and musical styles. But you know, without hesitation, when you're in a Christian environment.

However, the visual elements get you only so far. Powerful icons are visual shortcuts for something much deeper. And the icons of Catholicism™ are steeped in meaning and symbolism.

The crucifix represents the suffering of the Savior as he took on the sins of mankind. The bread and wine of Communion represent the Savior's body and blood that he gave up for believers. The act of baptism represents the cleansing of our sins.[7] These elements, and so many more, hold massive emotional significance for those who believe.

These icons and elements were then turned into products and have been emblazoned on merchandise for centuries, from leather-bound Scriptures to rosary beads and spiritual art. They've even been extended to more contemporary items like T-shirts and bumper stickers.

Catholicism™ continues to be a massive power in the world, and guides the actions of millions of people on a daily basis. The Catholic Church has done something right to continually enhance its **Staying Power** for over two millennia. That's something the board of directors of Coca-Cola can only dream about.

It's Hard to Argue with Numbers

We've covered plenty of psychological and emotional reasons in support of Iconic Advantage. But this is business—so we know you'll want to see some hard stats too. Fortunately, there's plenty of evidence that Iconic Advantage leads to greater customer loyalty, greater demand, and stronger testimonials.

Some research from WPP, one of the world's biggest advertising agency groups, shows that people find it far easier to remember iconic brands. Its studies have revealed that iconic brands have over 60 percent better top-of-mind awareness than non-iconic brands. This is because iconic brands connect more deeply with their audience. And they do it in a way that includes the heart as well as the head.[8]

Many marketers assume this connection is the result of frequently bombarding people with sales messages. But it seems that's just not the case. Research involving 7,000 consumers, conducted by CEB Inc., formerly Corporate Executive Board, shows that out of all people who claimed to have a relationship with a brand, 64 percent stated that "shared values" was the primary reason.[9] This is part of the very foundation of Iconic Advantage.

CEB's research went on to investigate what led to consumers buying a product. And it found that "making a purchase decision easy is what makes customers choose your brand." This is one of the results of building Iconic Advantage. By becoming the trusted product for a defined audience, you make your

Iconic properties deliver up to three times the profit and double the volume of other properties.

product the simplest choice when people are ready to purchase.

This deeper connection and increased awareness don't translate just into sales. They translate into more profitable sales too. Our experience has shown us that iconic franchises generate a disproportionate share of profits in comparison to noniconic products.

In our research, we looked at two dozen brands that weren't realizing their full iconic potential. Their iconic properties were delivering up to three times the profit and double the volume of other properties. That made them substantially more valuable to the business. However, the brands weren't realizing the full iconic potential of these properties because they weren't investing proportionally in line with their value. The companies were just taking the success for granted. If they allocated a bigger share of the budget to develop these iconic franchises, they would be able to carve out a larger share of the market, reach more people, develop product extensions, and do other activities that would generate a significantly bigger profit.[10]

Higher Sales and Lower Costs

Since the 1980s, corporations have been using Michael Porter's generic strategies to guide their businesses. At the root of his thinking is the belief that you are focused either on differentiation or cost leadership. You offer either a uniquely desirable product or a no-frills bargain.[11] There is no in-between. However, with an Iconic Advantage strategy, these options aren't mutually exclusive. And, indeed, they shouldn't be.

An Iconic Advantage approach helps you drive greater differentiation by creating powerful signature elements. This, in turn, generates greater loyalty and demand—which naturally translate into higher prices and greater sales volume.

This simply grows over time as you strengthen your differentiation by continually innovating your existing product. Your regular refreshes stimulate repeat purchase from your existing audience. And by bringing your product to the front of your audience's minds, you can spend less on marketing and promotion.

At the same time, the increased volume you enjoy from being a market leader drives down costs and allows you to achieve the highest gross margins in the industry. This, in turn, allows you to reinvest those profits into creating even greater Iconic Advantage. It's a virtuous cycle.

Innovating your existing products leverages your existing staff, structures, processes, manufacturing facilities, distribution channels, retail outlets, and customer base in a way that "radical" innovation never can. It's a smart use of resources and energy. It leads to increased gross margins and has a far higher chance of success.

It's also simpler to do, and you'll start seeing results faster.

Michael Porter's mutually exclusive strategies have had their day. You don't have to choose differentiation *or* cost leadership. You can actually have your cake and eat it too.

This is a smart and profitable strategy. That's why the brands dearest to our hearts focus on creating iconic products and commit to their development.

Having an iconic product is a conscious business decision, not a lucky break. It's just amazing that so few companies have embraced Iconic Advantage as a path to success.

Building Iconic Advantage

Developing Iconic Advantage isn't complicated. It just takes a bit of work. The next three chapters will explore each of the three steps in more detail.

Iconic Advantage Framework™

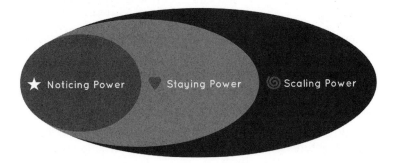

Chapter 3: Creating Noticing Power

We'll explain how to make your product **meaningfully distinctive** so that it stands out in the market.

Chapter 4: Developing Staying Power

We'll explain how to build a deeper connection with your audience by keeping your product **relevant over time**.

Chapter 5: Building Scaling Power

We'll explain the opportunities to leverage your **Iconic Brand Language**™ to build a larger following.

As you can see, it's not a complicated strategy. It's just amazing how few companies have harnessed it to get an edge on their competitors.

Let us show you how to get started.

CHAPTER 3

Creating Noticing Power

Few companies succeed in creating a product that can stand the test of time. So it's impressive how Nike—a company at the center of one of the world's most faddish industries—has managed to get nearly thirty years of mileage out of its iconic Air Max sneakers.

The original idea of air-cushioned soles was developed in the 1970s by M. Frank Rudy, a former NASA engineer. The technology was truly space age. The blow-rubber-molding technique had been developed for the space helmets used by Apollo astronauts. Rudy saw the potential benefit of using the technique to create air pouches for the soles of sneakers. The idea was that the cushioning would be more effective and last longer than that of standard sports shoes. Traditional foam-soled sneakers lost about 40 percent of their cushioning over their lifespan, but a pouch of air would never lose its bounce.[1]

Every sports shoe manufacturer had turned him down. They saw it as a costly, unproven technology. But when he took it to Nike, the company saw the potential right away.[2]

When Nike first built the technology into its Tailwind shoe, it embedded the air cushion within the structure of the sole. You couldn't see any difference. The only way people would know about the technology was through the marketing. And that made it feel like more of a gimmick.

The Nike Air Tailwind compared to the Nike Air Max 1.

Nike kept innovating on the idea.

The real stroke of genius came when Nike released the Air Max 1 in 1987. This shoe had a much larger air pocket and a little window into the air cushion on each side of the heel. The air pocket wasn't just visible; you could poke it with your finger and feel the cushioning effect. It felt futuristic. It looked sexy. And it sparked conversations. It was an immediate hit. And in the decade to come, it would earn Nike billions of dollars.

The bubble was a truly distinctive signature element. There was nothing else on the market like this. It was a visible and striking part of the shoe, giving it **Noticing Power** in spades.

On top of this, the design element visually communicated a mean-ingful benefit. The idea of running on air is pretty easy to get, and this design showed that at a glance. It wasn't just an aesthetic flourish—it

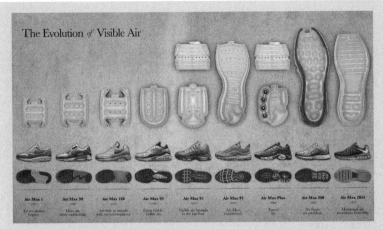

The Air Max idea had all the ingredients of an iconic design.

was a visual demonstration of the product's benefit. And its clear relevance greatly enhanced the brand's **Staying Power.**

The Nike air bubble looked incredible and demanded to be stroked, prodded, inspected, and admired. We're sure many people had a sniff when they thought no one was watching. This kind of design requires an understanding of what makes a product distinctive—and continued investment in that distinction to keep it relevant. It requires the right balance of old and new, familiar and fresh, heritage and evolution. It involves experimenting, failing, and persevering. It needs retooling and a redesigned manufacturing process. But that kind of investment is tiny in comparison to the ongoing sales such a design can generate as it remains relevant over time.

That's the power of great design. It attracts attention, it communicates a benefit, and it looks gorgeous. And if you do it as well as Nike did, it leads to profit—lots and lots of profit, year after year.

Nike succeeded in taking technology they already had and turning it into something iconic. This was a feature that had marketing built into it. It drew people into stores to see the Air Max sneakers and became a talking point for anyone wearing a pair.

This kind of approach is far more powerful than spending millions on an advertising campaign or a celebrity endorsement. And it gave Nike a property that it has been able to refresh and expand through the years. There can't be many business investments that have paid off quite as well as this.

Most manufacturing industries are on a runaway train heading towards faster, cheaper, and more commoditized products. It's all about quantity, quantity, quantity—which pushes quality into fourth place. Because why would you need to create a jacket that can last years when it's likely to be replaced by another one in a matter of months?

The companies that embrace this trend are all focused on what's next. They're concerned only with the new. But being iconic is about taking a step back and looking at the now and the then. It's often about innovating on your oldness to create something that rises above the accelerating speed of consumption. It allows a company to take an existing asset they've already invested heavily in and develop it to make it more distinctive, relevant, and universal. Makes business sense, doesn't it? That's the foundation of Iconic Advantage®.

What Exactly Is Noticing Power?

Just like the name suggests, Noticing Power is the ability of a brand or product to grab people's attention when lined up against the competition. It goes beyond rational and practical design to add an element of desirability, uniqueness, and distinction. This creates an emotional connection with the consumer, makes the product more memorable, and leads to lots more sales.

Most brands miss this opportunity spectacularly. They appear
to be more interested in standing
in than standing out. They have
the misconception that confor-
mity and blandness are the routes
to success. They think that if the
competition is doing it one way,
it's probably a good idea to follow

> *It's about innovating on your oldness to create something that rises above the accelerating speed of consumption.*

suit. They make the mistake of believing that the iconic brands
got their success by accident rather than intent.

And that's great news for you.

The lack of vision intrinsic in most brands makes the opportunity so much bigger for the visionary brands that embrace
distinction.

This chapter is all about creating elements that help a brand
stand out in the market and stick in people's minds.

The Secret of Noticing Power

We live in an age of bland homogeneity. Most markets have
become a race to the middle in which everyone loses out.

It's often difficult to distinguish one midsize car from
another. If you cover the emblems on a Toyota Camry and a
Chevrolet Malibu, you'll have a hard time telling them apart. And
with their sets of almost identical features, the only reason to
choose one over the other is the price.

The finance industry has a similarly uninspiring cookie-cut-
ter approach. Banks all offer subtle variations of the same thing,
delivered with comparable service by identical-looking staff.
They try to differentiate themselves by offering a fraction of a
percent better interest rate or by launching a smartwatch app a

few weeks before everyone else does. These activities can never have any long-term impact on the business.

Most brands...make the mistake of believing that the iconic brands got their success by accident rather than intent.

Even most websites follow the same look, taking their lead from Amazon or Apple or whoever the market leader is in their industry.

Hiding in the crowd doesn't get you noticed. The good news is, in a world of corporate sameness, it's really easy to stand out.

However, it's not just about standing out for the sake of it.

Products need to be noticed for the right reason. Your Noticing Power needs to be built on a truth. And that truth needs to come from an understanding of your business.

What Makes You Unique?

There are different facets to what makes a business and its products special. It's surprising how few companies understand these important details. Instead, they just wing it without any real understanding of themselves and how they're perceived. We don't recommend you do that.

Completing an Iconic Brand Pyramid™ is a simple way to understand more about who you are. We cover this in more detail in Chapter 6. But here's a brief overview so that you can see where your signature elements fit in.

Iconic Brand Pyramid™

Start with the foundation. Make sure you understand your purpose and values. This is the unifying belief that motivates your employees. This is their reason for doing what they do. And it's central to the culture of your organization.

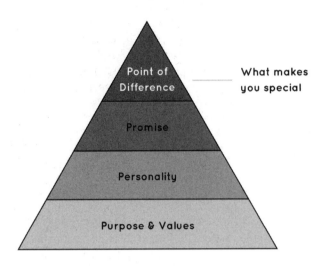

Next, you need to get a clear understanding of your brand's personality. This is what drives your organization itself. It's the character that influences all of your activities, communications, and decisions as a business.

Then you need to make sure you have a well-defined promise. This is the publicly visible part of your Iconic Brand Pyramid. It's informed by all the other levels. And the more defined it is, the better your audience will understand who you are and how you can benefit them.

The pinnacle of your Iconic Brand Pyramid is your point of difference. It should fit seamlessly on top of the layers you've just defined. It should help you deliver the promise and be grounded in your personality, purpose, and values. This is where your signature elements fit in. Ideally, they'll be an embodiment of your key point of difference.

For example, Nike's air pocket embodies their point of difference of improved performance, which in turn is a manifestation of their promise to empower their customers to "just do it." The better your signature element helps you to deliver your promise, the better it will be in building your Iconic Advantage.

Defining Your Signature Elements

Products with real Iconic Advantage have one or more signature elements that capture their audience's attention and hearts. These elements become a symbolic representation of the brand. They can't just be something generic that the competition has; they need to be unique to the product and communicate something about what makes it different.

These elements must be distinctive, relevant, and universal. You need all three of these facets to get the most benefit. This is the very foundation of Iconic Advantage.

▶ **Be distinctive**

The most obvious way of being distinctive is visually. You want your product to stand out when it's up against the competition. And usually, that's not difficult. Most categories are overflowing with generic products. But distinctiveness can involve any sense. You can have a distinctive feel, smell, taste, and sound too. And you should also consider having a distinctive point of view. Most brands are scared to have an opinion for fear of alienating potential customers, so this can be a powerful opportunity for the right brand. Ultimately, the brands with the most memorable distinction are able to create signature experiences and moments.

▶ **Be relevant**

Once you've caught people's eyes, you need to grab their hearts and minds. This is about being noticed for the right reason. If what makes your product different is an obvious benefit to your audience, you've hit the nail on the head. The original iPod looked simple in comparison to all the other mp3 players around. Competitive products were covered in buttons and had fiddly menu systems. The iPod was as elegant to use as its design suggested. It's this thinking that will make your product desirable.

> *Iconic Advantage is about building in the desirability at the start rather than trying to tack it on at the end.*

▶ **Be universal**

Now that you've got something that makes your product different for the right reasons, you need to make sure your product gets in front of as many eyeballs as possible. You need to make it universal. That involves a distribution strategy to get it in front of the right people at the right time as well as a marketing strategy to increase awareness.

In summary, you need to create signature elements that are noticeable, meaningful, and memorable.

Invest in the Right Place

Normally a brand's resources are focused on the marketing side of the business. Only after a product has been created does the business start thinking about how to make people want it. An Iconic Advantage strategy is the opposite of that.

Iconic Advantage is about building in the desirability at the start rather than trying to tack it on at the end. Which means that a smaller amount of marketing spend goes further.

Signature Elements in the Real World

We've already established that signature elements are notice-able details that make you stand out from the competition. These work together to form an Iconic Brand Language™. This is a combination of key signature elements aesthetically designed to create an overall look. This style is also something that can stand out and add memorability.

As an example, let's look at the classic Converse shoe. It has a number of key signature elements that haven't changed in decades. These include the circular logo on the ankle, the rubber sole and toe cap, the metal eyelets for the laces, and the lace guide on the tongue. These elements haven't lasted so long out of a lack of imagination—they've stuck around because they're identifiable. These elements combine to create the Iconic Brand Language of a Chuck that is universally recognized. Later in this book, you'll find out the effort Converse put in to update the shoes without losing their distinctive style.

Let's now look at BMW. There are a number of elements, besides its emblem, that are pretty consistent across their cars. Visually, these include the kidney-shaped grille, the contour lines on the hood of the car, and the Hofmeister Kink on the rear windows. But BMW also pays very careful attention to what their cars sound like. In fact, it has a team of sound designers at its Research and Innovation Centre in Munich. They pay careful attention to every noise a car makes, from the purr of the engine to the clunk of a car door closing to the satisfying click of a button.[3] It may seem a bit excessive, but these sounds communicate how well-built the cars are. The cars have a distinctive sound that communicates a relevant message that is universally applied across BMW's product line. Sometimes being iconic goes beyond what you can see.

Humans Are Built for Icons

Humans are naturally drawn to icons and their signature elements. Icons are the very building blocks of culture, religion, and mythology. And it appears that our brains are actually wired for them. A recent MIT study found that human brains

can actually recognize concepts within images in as little
as 13 milliseconds.[4] Our visual
cortex appears to be a lot more
powerful than we previously
thought.

Our recent understandings
of the brain show us that there
are some inherent processes
that we can take advantage
of to grab people's attention.
Our brains are programmed to
direct our attention towards
a certain kind of input. This is
called preattentive processing. It directs our eyes, ears,
and other senses to particular
stimuli before we even become
consciously aware of them.
Visually, we're drawn to certain
triggers, including strong
edges, intense brightness, and
areas of high contrast. We're

also irresistibly drawn to faces
and can even recognize emotional expressions before our
consciousness kicks in.[5]

**Flickr images used with
permission.**[7]

It's no surprise that faces—and the emotions they express—
are so important to humans. As we're social creatures, it's useful
to understand the state of mind of those around us if we want to
bond, mate, or keep ourselves out of danger. We're so strongly
wired to focus on faces that we even register inanimate objects
as faces if they have the right dimensions. This tendency to

perceive meaning in a random pattern is called pareidolia. If you've ever seen a face in a wallpaper pattern or a coffee stain or an electrical socket, you've experienced it.[6]

Understanding how our brains are built to process visual input can be useful in creating attention-grabbing signature elements. Because if elements are not immediately noticeable, they're not working hard enough for you.

Be the Product That Springs to Mind

Another principle that shows the power of iconic design is prototype theory. Research shows that people don't look at items within a category as being equal. Some are more "prototypical" than others. For example, research shows that when asked to give an example of an item of furniture, people are more likely to answer "chair" than "stool" or "drinks cabinet." We create mental hierarchies within categories.[8] And items with the most memorable signature elements make it to the tops of those lists.

This applies to brand categories too. Let's try a little experiment. Take a moment with each of these:

Think of a car.
Now think of a computer.
And finally, think of a next-day delivery service.

Chances are, you didn't think of a generic representation of a car, a computer, or a delivery service—you thought of specific examples. These are the *prototypical* examples of these products or services for you. The aim of Iconic Advantage is to give your product a better chance of becoming prototypical for your target audience. That means your product springs to mind more often when people are considering a purchase. It's like an advertising billboard in exactly the right place and at exactly the right time.

That's media space you just can't buy.

Designing Signature Elements

So how exactly do you design for signature elements? How do you create something that will earn you a greater share of attention?

A lot of that comes down to understanding your audience and the competitive landscape. You should be looking to offer the consumer something of value that no one else is offering.

It's not just about removing the pain for the consumer. Yes, that's a great place to start. But if you want to maximize your chance of Iconic Advantage, you really need to go beyond problem-solving: you need to look at adding pleasure.

It's important that you make people feel something. If you give users a little frisson of joy when they purchase or use your product, they're going to want to do it again. And again. And they're going to tell other people about it. That's where you have the real power over your generic competitors.

Iconic Identity

Most businesses understand the importance of their products' having a unique identity. That's why they give them a name and logo. You don't want your products to get muddled up with competitors' products, after all.

The best companies have brand guidelines that keep people on track and protect against brand dilution.

If your business does that, you're off to a great start. But it's just that: a start. This is only table stakes. You must go beyond this basic branding level if you want to really build your **Iconic Advantage**.

Iconic Feature

Offering something tangible that no one else offers is a great way of standing out. The first step is to make sure this feature is prominent and visibly (or experientially) noticeable. But you need to make sure that it goes beyond just being noticed to demonstrating a real benefit. It could be a more elegant process, some unique technology, a special ingredient, a functional design element, or any number of other things. For a product to be successful, it needs to benefit the audience.

The PC market in the late 1990s and early 2000s fell into the trap of competing on meaningless numbers—like RAM, processor speed, memory, and graphics card version. It became impossible to compare PCs on a like-for-like basis, because very few people understood what each of the numbers meant for them. Is more RAM better than greater memory when it comes to writing a Word doc? Does a 32x CD-ROM drive help someone play games? And while the market battled over meaningless mathematics, Apple came in with emotional benefits and captured customers looking for a computer that represented who they were.

If you can offer a relevant benefit that no one else does, in a way that's hard to miss, you're well on your way to success.

James Dyson is your classic British inventor. He created over 5,000 prototypes on his way to perfecting the dual cyclone vacuum cleaner. After every big manufacturer had turned down his revolutionary idea, he decided to go into business for himself. He had a unique feature in the market, but he knew that the benefits are what would sell his product, rather than the technology. He used a clear collection bucket to visibly show that his vacuum cleaner was bagless and therefore never lost suction power. His vacuum is now sold in over 65 countries and is regularly refreshed with updated technology, different sizes, and new designs.[9]

Iconic style

In many markets, what you need is a distinctive look. You need something that is ownable and different. Adding an emotional story or a heritage element to this design element helps it connect even more deeply with the consumer.

The famous Burberry check pattern and beige color scheme are recognizable at a glance. The brand has an amazing 150-year history, having outfitted polar explorers, movie stars, and American presidents. The classic and timeless look of its core garments adds to this feeling of heritage. Wearing one of Burberry's products connects you with that history and the stylish icons who bought into the brand before you.

Iconic silhouette

Having a distinctive product shape is a great way of standing out in a sea of sameness. Take a look at the shampoo shelf of your local supermarket and you'll see that most products look pretty similar. You'll notice a mass of generic-looking plastic bottles with flip-top caps. It doesn't take much to stand out in that environment. The classic Coca-Cola bottle is probably the best-known example of a distinctive silhouette. But you can probably think of examples in most sectors, from cars to phones to footwear.

The iconic Kikkoman bottle was launched in 1961 and hasn't changed since. The iconic teardrop shape is topped with a smart red nondrip cap that is both functional and distinctive. Kikkoman gives the bottle's design the credit for its success, saying that it took the product "out of the kitchen into restaurants and dining rooms."[10] The condiment was no longer just a drab ingredient. The stylish bottle turned it into a condiment that deserved pride of place on the dinner table.

Iconic Experience

So many companies miss out on the power of experience. We're not just talking about creating an event here. Instead, we're talking about how you can make users feel something when they interact with the product. It can be as simple as creating witty feedback messages on an app or showing people a satisfying way of enjoying your product (like Oreo's twist, lick, and dunk approach). Create a unique experience and you'll stand out from your monotonous counterparts.

When people talk about Apple, they tend to focus on its products. But one of the things Apple does better than just about any other company is its packaging. When you unbox a new iPhone, you get several layers of delight before you even see the phone. Then once you've removed those, you're presented with some beautifully packaged headphones and a stylish power plug. It's a gradual unveiling that's designed to heighten the experience of buying one of Apple's products. This is a missed opportunity for thousands of brands.

In the 1980s, Richard Branson, the founder of Virgin Records, decided to start his own airline, Virgin Atlantic. People thought he was crazy. He was a music business guy who knew nothing about the airline industry. But he understood what it was like being a passenger. And he knew it was often a disappointing and frustrating experience. So he set out to create a better experience. Virgin Atlantic offered stand-up bars in upper class, limo transfers, massages, and beauty treatments, and developed seatback video long before other airlines.[11] These innovative experiences, along with the attitude of the flight crew, helped the company quickly become iconic in a ruthless industry—because it understood that the journey is just as important as the destination.

Iconic Sensory Stimuli

Don't just aim for the eyes. You can create an iconic element with any sense. Fashion is as much about feel as it is about looks. Food is as much about appearance, texture, and aroma as it is about taste. And, as you've already read, cars can benefit from sound design as well as visual styling. Think beyond the obvious and try to build something iconic by appealing to another sense.

When you ride a Harley-Davidson motorcycle, you're riding an icon. It's the pioneer spirit of the American West and a powerful symbol of freedom. But when you come across one on the street, you probably hear it before you see it. The growling *potato-potato-potato* of its V-twin engine is so distinctive, the company spent years trying to protect it. A Harley's ribcage-rattling rumble is the very essence of the product.

Iconic Spokesperson

All over the world, you'll find real and fictional individuals attached to products. George Clooney is the sophisticated face of Nespresso coffee. Tony the Tiger has been adding fun to Kellogg's products for over sixty years. Soccer legend Pelé tackled the embarrassing topic of erectile dysfunction. It's a tried-and-tested route that really works if you get it right. But it involves getting a spokesperson who's more than just a familiar face. This spokesperson needs to help you communicate something about your product or brand. That could be aligning with your brand values, embodying the benefit, illustrating the problem, lending an air of trustworthiness, or something else. If the person doesn't add something relevant, use someone else.

Allstate's advertising agency, Leo Burnett, took an interesting approach with Allstate's spokesperson. In Allstate ads, Dean Winter plays the role of Mayhem. He's the personification of all the ridiculous things you should be insured against. In one ad he plays the part of a teenaged Valley Girl who loses control of her pink SUV because of an upsetting text. In another, he's a raccoon cavorting and breeding in someone's attic. Whatever the ridiculous problem is, the ad ends with a strong sales message about how Allstate can insure you against it. The campaign has proved to be as effective at winning customers as it has at winning advertising awards.[12] All thanks to a memorable and unconventional approach in choosing a spokesperson.

Iconic Point of View

Most companies are so afraid of alienating any potential customers that they won't express any opinions. But trying to be all things to all people tends to make you nothing to anyone. You won't engender passion in your audience if you don't show any yourself. In recent years, a few notable brands have realized this and have stood for something bigger than themselves. Unilever has tried to spread this thinking across all of its brands. This has resulted in Dove's Campaign for Real Beauty, Omo's "Dirt Is Good," and Domestos' sanitation program.

But it's not just about being socially responsible. It's also good to stand for something that will generate debate. You may lose a few friends along the way, but you'll create a level of passion in your customers that you can never achieve with quiet timidity.

When Red Bull launched in 1987, it created a whole new category of soft drinks. But more important, it created a new mindset for adventurous sportspeople. At the center of everything the company did—from selling sugar water to supporting extreme sports—was the belief that it was about enhancing people's mental and physical performance. Red Bull started sponsoring sports events and even seemed to create thrilling new ways to cheat death. Its wingsuit team members merrily throw themselves off mountains without parachutes. Its surf team tackles giant waves that were previously inaccessible. And in 2014 the Red Bull Stratos jump involved Felix Baumgartner's diving to earth from the edge of space. But opening up one of Red Bull's cans is exciting enough for most people.

Iconic Name

Trying to be all things to all people tends to make you nothing to anyone.

These days a name like Smithers, Boggins & Dunwoody Inc. just doesn't cut the mustard. And when you do a web search, it seems like all the most useful terms you could pick for a brand are already gone. So the challenge is to find a name that is distinct and memorable and communicates something—even if that's just a feeling. If you look at the most iconic digital brands from recent years, you'll notice that a lack of obvious words hasn't held them back. Some are descriptive, like Netflix (movies via the internet). Others tell a story, like Airbnb (the founders originally bought three airbeds and rented them out as an "air bed-and-breakfast"). Some are a statement of ambition, like Uber (German for "an outstanding thing"). And others are just expressive terms, like Shazam. Whatever approach you take, you need to make sure you don't get muddled up with another brand (for practical and legal reasons).

In 1932, Ole Kirk Christiansen founded The LEGO Group in Denmark. The name is a contraction of the Danish words *"leg godt,"* which mean "play well." The company started making children's toys, beginning with a wooden duck. It didn't release its iconic click-together blocks until 1958. And after that, it discovered that *"lego"* is actually Latin for *"I put together"*—an amazing coincidence.[13] But the simplicity and uniqueness of the name are what led to its ease of international adoption.

Creating Noticing Power

At this point, you're probably asking how you can take this thinking and start applying it to your business. You may have a brand or product that you want to refresh. Or you may be looking at ways to give a new product the best chance of success. These require quite different approaches, so let's deal with them one at a time.

If you already have a product, service, or brand that could do with a boost, your job is to find the existing elements with the best potential of becoming iconic. This isn't something to do alone at your desk; it's something you want as many people's input on as you can get. Ask colleagues from across the organization to list all the elements for the property you're dealing with. Ask them specifically to look for elements that are unique to your offering. It's probably not a good idea to turn a generic feature into your iconic hook.

If you're dealing with something that's entirely generic, with nothing unique or ownable, you'll need to deal with it like a new product.

With something new to the world, there is an opportunity to develop first-mover advantage (FMA). Being the first to market with a new kind of product is one of the best ways of creating Noticing Power. Just think of classic brands like Kleenex, Clorox, and Q-tips that were early—or even first—to market with their products. That put them in an enviable position.

If you succeed at this, you deserve a pat on the back. Now you just need to avoid the slap in the face of losing that advantage. You do that by making sure you've got strong signature elements that will help your product remain the best and most distinctive option as other brands follow in your exciting footsteps—because being first means nothing if someone else becomes the best.

Next, you need to focus on developing lasting relevance by creating Staying Power. We'll tell you more about that in the next chapter.

The first-mover advantage is also a great approach if you're creating a brand or product from scratch. Again, invest some effort in developing your signature elements. It's vital that you do this well from the get-go, so you'll need to start with research and analysis. Look at the competition and list their signature elements (if they have any). Plot them on a matrix, sort them in different ways, and play with them until you find some gaps. Then start generating your own signature elements that can fill those gaps.

We look at how to do this more fully in Chapter 7 when we show you how to create an Iconic Brand Language guide.

Let's Start Getting Noticed

Most of this chapter probably sounds pretty obvious. Of course, standing out from the competition is the best way to get noticed and attract purchasers. However, take a stroll up and down a few aisles in your local supermarket and you'll see that hardly any company follows this advice. The "unique benefits" they offer are usually not unique at all and are often entirely fabricated pseudoscience (yes, cosmetics industry, we're looking at you).

To build Iconic Advantage, you need to start by creating a property with desirability built in, rather than expecting marketing to make up for the shortfall.

In the next chapter, we'll explain how to create timeless relevance for your signature elements, something that requires as much heart as it does head.

Iconic Advantage Toolbox

It's impossible to properly assess a brand or product's Noticing Power if it's part of your job. You just can't see it the way your audience does. So here are some exercises to help you find out if you really have the **Noticing Power** you need.

Do the Double-Take Test

Put your product somewhere visible—like at the front of a store window or a display stand—and watch the reactions of pass-ersby. Does it catch their eye? Does it make them do a double take? How many people notice it? Do they have anything in common? You can do this informally to get a feel for people's reactions. And—if you want to quantify the results—you could measure all of these metrics and use them as a benchmark to test improvements against.

Do the One-Second Test

It's good to get an understanding of what people will recall about your product. Get some test subjects and flash your iconic signature elements for just one second each. You can also mix these stimuli with your competitor's signature elements to get a true feel for how you compare. Then test the recall of your subjects. If your visual elements are iconic enough, they'll be remembered. You can also run this test with the other senses if your signature elements aren't just visual.

Draw the Product

Hold one-on-one sessions with consumers. Briefly show them the product and then ask them to draw it. It doesn't have to be a quality sketch. A low-quality scribble with descriptive labels is just fine. The subjects will naturally draw the parts they remember. After you've done this a number of times, you'll be able to see what people find important and what they leave out.

Do the Signature Moment Test

Give people an experience of your product or service. Then ask them what moment stood out for them most—and why. Is it the experience you wanted to leave them with? Is it a frustration that you maybe need to deal with? Is it another moment that might open up a new opportunity for you? If you want to get a better understanding of the competition, do it for competitors' products too.

CHAPTER 4

Developing Staying Power

The internet of the 1990s was a very different space. It was just finding its way, and many of the digital things we all now take for granted simply didn't exist. One of these things was how we found information.

The big innovation that took the internet from the land of the übergeek to the mass market was the link: the ability to click on a button, an image, or a piece of text and be taken directly to another page. It seems inconceivable now, but before the birth of the hyperlink, you needed to type in the entire address of each webpage you wanted to visit. And the web wasn't nearly as easy on the eye as it is now.

For the first few years of the web, the only way to find the sites that were of interest to you was through a directory system. You'd click on the link for "automotive" and then the link for "sports cars" and then the link for "Lamborghini," and eventually you'd make it to Lamborghini's website. You'd find these directories on pages known as portals—which were usually the first page that loaded when you opened your browser.

These portals all included a search box as well. But their search features were rather hit and miss. If they'd indexed a webpage that mentioned Coca-Cola more often than Coke's own page did, that webpage would usually appear higher in the list of search results.

So in the mid-'90s, a couple of guys at Stanford University, Larry Page and Sergey Brin, began working on a better way of indexing websites by looking at authority, relevance, and freshness. The results their algorithm produced were significantly better than those from the search engines embedded in the portals. When Page and Brin launched Google officially in 1998, the search function was all they had developed, so that was all they could put on the page. They had no advertisers, no news service, and no categorized database of websites to add. Their page ended up rather sparse in comparison to the busy and cluttered portals that were the norm. And that played to their advantage.[1]

Looking entirely unlike a portal got Google's page noticed. The sparse layout of the page drew people to its point of difference and made it distinctive. Google quickly got a reputation as the best search engine on the web. The portals suddenly found themselves obsolete.

Over the years, the overall look and feel of the Google homepage hasn't changed much. But this spartan consistency has managed to stay fresh and relevant while the portals vanished into history.

From the very start, you could always recognize the familiar Google homepage for its simplicity and the prominence of the central search box. This consistency in look was matched by a consistency in behavior, wherein the logo changed on an almost daily basis to mark occasions or celebrate special people.

Importantly, Page and Brin also stayed true to their mission of democratizing information, even while they publicly dealt with the

Excite[4] and Google homepages. (Google and the Google logo are registered trademarks of Google Inc., used with permission.)

growing pains of Google's becoming a business behemoth. They continued to pursue new search innovations, including Google Maps, Google Earth, YouTube, and Google Play.[2] This ability to stay familiar while evolving Google's story, innovating its key benefit, and keeping the site design fresh has allowed the company to stay relevant all these years.

And all of this has led to its attracting over 80 percent of the search traffic in the US.[3] It's true. Just google it.

It seems that some brands are just better at sticking around than others. If you ask businesspeople why that is, you'll get a variety of answers.

Many people think it comes down to sheer luck. Admittedly, success always has an element of someone's being in the right place at the right time, but you tend to find that the people who benefit from an opportunity have put themselves in that place at that time intentionally. Luck isn't a strategy. And assigning a company's success to an alignment of the stars is ignoring the thinking that helped it grab an opportunity before anyone else.

Some people might think that the credit goes to a bottomless marketing budget. But as an old advertising adage goes, the

worst thing you can do is advertise a bad product well. People will
try the product once, realize how
awful it is, and vow never to waste
their money on that thing again. If
you don't have the right product,

The worst thing you can do is advertise a bad product well.

no amount of advertising and promotion can turn it into a long-
term success.

Another myth is that longevity comes down to strong leadership. But unless that strong leadership is driving the company
in the direction of making products that stand out and stay relevant, no amount of boardroom magic can convince a market to
choose your product over the competition.

This fallacy also applies to the idea that success comes
down to rock star talent. But it's more than just hiring amazing
people; it's giving them the environment to thrive in. Even mod-
erately talented people can excel in a great environment, while
phenomenally talented people will be stifled in a bad one. If an
organization has the right conditions for people to flourish, it
will continue to prosper when the time comes for its biggest
rock star to leave the building.

Finally, many people believe it's just about the ability to spot
a trend and ride it into the sunset. You can do quite well with
this strategy but only in the short term. It can help you create a
fad or a craze, but it's unlikely to earn you any sustained, long-
term success. The best businesses stick around. They don't only
weather the downturns; they continue to pick up market share
in the process.

So if none of these is the reason for the ongoing success of
a product, what is?

We call it Staying Power. Google has it and so do many
others. We'll show you how to create it in this chapter.

ATIC

Moving Beyond Noticing Power

In the last chapter, we showed you that Noticing Power is all about getting a bigger share of attention. It's about using signature elements to differentiate your product from alternatives. It's about being meaningfully different to attract a bigger share of attention, which in turn leads to more sales. The trick is sustaining and enhancing that advantage over time.

That's what we call Staying Power. It's about sticking around to become part of your audience's lives. Even in a small way.

You can do that only by being relevant both emotionally and rationally.

Over time, consistent relevance builds a meaningful connection. And just like in personal relationships, the amount of time you spend with a brand matters. Studies show that familiar brands can end up having connections as strong as those we make with our closest family members.

Professor Gemma Calvert, founder of the neuromarketing agency Neurosense, conducted an experiment with the BBC to look into this. She placed people in an MRI scanner and showed them pictures of products—a Heinz beans can, a Coke can, a can of Red Bull, a McDonald's meal—followed by photos of their friends and families. The results were striking. They showed that we use the same brain areas to recognize both familiar brands and familiar people.

These familiar brands even cause similar sentiments to those we experience when we see a loved one. That's the result of repeated exposure to consistent elements.[5]

Connecting on both rational and emotional levels may sound unattainable at first—and possibly a bit on the fluffy side to the more cynical amongst you. But don't be fooled; the companies behind the most successful products understand the importance of building a lasting connection. And they do it by ensuring they remain relevant over time.

The Dimensions of Relevance

For a product to maximize its relevance, it needs to address two dimensions: *connection* and *time*.

You can't just expect a sudden explosion of success to continue at the same level month after month, year after year. People very quickly tire of even the most novel and amazing products. If you're old enough, you'll remember how amazing the internet was when dial-up modems became a thing in the early '90s. Now think about how frustrated you feel when a specific episode of an obscure TV show doesn't immediately stream on your smartphone. Internet access went from an exciting novelty to a public utility, like water and electricity. The sparkly shine of novelty tarnishes at a frightening rate, and formerly jaw-dropping innovations become the norm shortly after they hit peak adoption.

Simply focusing on satisfying rational needs with ever-greater functionality becomes a pointless rat-on-a-wheel cycle. To break out of this, you need to move into deeper emotional territory where loyalty remains, even against competitors with bigger bells and whistles. It isn't always the biggest and baddest mousetrap that wins. Often it's the stinkiest cheese that gets the mouse.

So you can't rely on a purely rational approach if you want long-term success.

Additionally, to become truly iconic, you need to stand the test of time. It's easier to have a place in people's futures if you're already grounded in their pasts.

If you're not built to last, you're likely to become a flash in the pan. Chris Anderson pointed out in his book *The Long Tail* that the really

> *It isn't always the biggest and baddest mousetrap that wins. Often it's the stinkiest cheese that gets the mouse.*

significant profits often come over time.[6] If you're nothing more than a fad, you won't benefit from that.

Let's look at these dimensions in a little more detail.

Rational ◄──────► Emotional

The first dimension is *connection*, split into rational and emotional.

Building a connection with people goes deeper than mere reason. Think about all the reasons you feel connected to other people: similar interests, common values, shared experiences, making you feel good about yourself, regular contact, mutual respect. All of these things apply to organizations wanting to connect with individuals too.

Emotional connection is important at every stage of audience engagement. But its value is most noticeable when it comes to buying the product.

It's been known for years that people tend to make purchase decisions on an emotional basis and then back up their choice with rational reasoning.[7] Yet most industries still try to convince people with facts and stats and well-reasoned arguments. Especially in the business-to-business arena.

We're not saying that you should concentrate on emotions alone. It's not an either/or situation. Instead, the rational and emotional approaches support and strengthen one another. We're saying you should cover both angles. By looking at both rational and emotional elements over time, you can develop a strategy to help your products stay the course and become relevant on a more sustainable level.

Past ◄──────► Present

The second dimension is time, split into past and present.

The past contains everything that has defined the product so far, from its conceptual and physical development through to its brand communications. It's important to understand that this includes your audience's historical understanding and experience of the product. This is the sum of every contact they have had with the product and brand as well as everything they've heard about it, true or not. It's vital that you con-

The ability to remain timelessly relevant allows you to step dynamically forward with one foot grounded in your past.

sider your audience's point of view if you want to truly connect with them.

The present is about making sure your product stays current, both now and in the future. Most companies treat product development like tectonic plate movements—they resist change until the pressure builds up to a level that something significant has to happen. These seismic shifts can be costly and harmful for the business, and they risk your alienating your existing audience. On the other hand, if you're keeping an eye on your continuing relevance, regular changes are part of your strategy. In that way, you evolve with your audience rather than take a leap into the unknown.

The ability to remain timelessly relevant allows you to step dynamically forward with one foot grounded in your past.

The Relevance Matrix

Bringing together these two dimensions gives us a matrix that you can use to guide your activities. Building on each of these four areas will help you build relevance and strengthen your connection with your audience. For the rest of the chapter we'll cover these areas one by one.

Iconic Relevance Matrix™

past ⟷ present

rational

emotional

| protect signature elements | innovate benefits |
| evolve story & heritage | reimagine design |

Protect Signature Elements to Maintain Familiarity

Have you ever bumped into someone you haven't seen for a long time and, even though they've changed, you still recognize familiar things about them? Like their smile, their accent, or the way they use their hands when they talk? Their clothing, hairstyle, and number of wrinkles may be entirely different, but the essence of what makes them who they are remains familiar. Successful products and brands also retain their familiar signature elements over time. And audiences tend to rise up against brands that meddle with these elements (like Gap did with its logo in 2010[8]), just like they react to celebrities who change their accent.

Consistent behavior makes things easier to understand. Think of traffic lights. In most countries, they consist of a three-colored sequence: red, yellow, and green. These colors don't change, so you don't need to spend any time considering what the signal means. This consistency leads to an automatic understanding of the situation without any conscious thought being required. That's the benefit of familiarity: it conveys meaning without any effort whatsoever. And from a psychological point of view, the fact that familiar objects, faces, and symbols are easier for our brains to process makes them preferable to new ones.[9]

Existing awareness is so important to us that it can skew our decision-making. One of the most cited biases in behavioral economics is the familiarity bias, which indicates that people tend to buy products they already know or have purchased before.[10]

A sociological theory of trust breaks this down further to state that familiarity leads to confidence and confidence leads to trust. And in many cases, trust is a precondition of purchase, particularly for higher-priced products.[11]

So it's easy to understand why the most popular products remain consistent, even as they evolve. Google's logo changes on a daily basis, but it's always positioned above the familiar search box and it maintains a spirit of creative playfulness.

You can see a rock-solid consistency with the Coca-Cola brand over time. It has evolved over the years, but it's never lost its Coke-ness.[12] This extends across the entire line of Coke's products. All the products use the same color palette, logo, and

That's the benefit of familiarity: it conveys meaning without any effort whatsoever.

fonts—or at least some combination that feels suitably familiar. Their ads all have a similar look.[13] If the font, logo, colors, and shape changed from bottle to bottle, you'd have a harder time trusting it—because the consistency of a brand signifies the consistency of the quality of the product.

Create an Iconic Brand Language™

To protect familiarity, we recommend defining an iconic language that sits at the core of everything the brand does. This shouldn't be just about external communications either—it should influence the internal workings of the company and its culture and be a guide for every department. That's the only way to create the kind of consistency that brands need to reach the iconic level.

But most brand guides don't go far enough. In Chapter 7 we'll provide a detailed guide on how to create your own Iconic Brand Language. For now, it's important to understand that there are assets beyond the basic branding guides that also need to be protected.

> ▶ **Signature Elements**
>
>   As we mentioned in the previous chapter, iconic properties need to have signature elements to stand out from the competition. These elements become stronger over time as your audience gets to know them and associate

them with your offering. As such, they become increasingly valuable and need to be protected.

▶ **Brand DNA**
Brand guides tend to focus on the way a brand portrays itself visually—and often also cover the tone of voice that a brand will use. But in today's world, where static media is on the decline, the brand's behavior is possibly even more important. It's important to understand the brand's promise, its personality, its functional and emotional benefits, its values and its purpose. These reveal the brand's motivations and can be used to guide decisions and behavior company-wide.

▶ **History and Heritage**
The histories of your product, brand, and business can be valuable assets to draw upon. They form the basis of stories that your audience can connect to—and demonstrate a consistent product journey that can be used to build trust.

As we've already said, iconic products play a special role in the lives of their audience. And consistency is an important element of that connection. Icons act as solid anchors, giving believers something to define themselves by. If your brand is wildly inconsistent across its touchpoints, it's unlikely to make it to the iconic level. Consistency leads to familiarity, familiarity leads to confidence, and confidence leads to trust. Trust, in turn, leads to more sales, more recommendations, and more profit. Consistency is essential if you want to symbolize something in your industry.

There are many approaches to creating a strong Iconic Brand Language. But once you've found the right approach, you need to stick with it for the long term.

Kit Kats used to be so much more than a chocolate treat. They were an experience. They were a ritual. And the enjoyment started way before you took your first bite.

The chocolate fingers were wrapped in a thin layer of tinfoil, then enveloped in a glossy red band of paper. The act of unwrapping them was a delight in itself.

Step one was sliding the silver block out of the paper band. If you wanted, you could smooth the pad of a finger down the length of the chocolate bar to reveal the imprint of the logo, like a child doing a brass rubbing with a crayon.

Step two was running your nail down the groove between two of the fingers. This would neatly slice the tinfoil wrapping, giving you a satisfying metallic crinkle as you did it.

Step three was separating the first chocolate finger with an audible snap, before slipping it out of the slice in the foil you'd just made.[14]

After that, you ate it.

This little ritual was one of Kit Kat's signature elements. It was shown in TV ads.[15] It gave pleasure to everyone who performed it. And then the makers got rid of it—for the individual packs, at least.

In the early 2000s, individual Kit Kats moved to a very standard plastic wrapper. The ritual was gone. And with it, a good amount of the pleasure the product offered. Since then, YouTube has revealed people's passion for unboxing and unwrapping.[16] The old-style Kit Kat bar belongs there.

This situation could have been avoided if the signature experience had been protected with a strong **Iconic Brand Language.**

Many other products provide little rituals for their audience that manufacturers are unaware of or don't care about. These are missed opportunities if the manufacturers let them go without understanding their power.

Evolve Your Story and Heritage to Create Meaning

People love to tell stories. If you spend $5 on a bottle of wine, you'll likely pour it into your guest's glasses without saying anything. If you spend $80 on a bottle of wine, you're probably going to mention something about where you first tasted it or who told you about it or how you once cycled through that region when you were a student. Stories are valuable. They add meaning, and meaning activates our emotions. This is key to building a stronger connection with your audience.

Humans just can't help adding meaning to objects. We do it pretty much as soon as we're aware of the world around us. Children often create a strong attachment to an object, like a blanket or a soft toy. And many children become fiercely protective of their own toys and become agitated when other kids try to play with them. We don't really grow out of this; it just evolves as we get older.

Studies have shown that, particularly in the West, we use objects to signal our identities. A 2011 study at the University of California, Berkeley coined the term "Prius effect" to describe how individuals will "undertake costly actions in order to signal their type as environmentally friendly or 'green.'"[17] This goes beyond conspicuous consumption, wherein individuals buy expensive objects to show off their wealth, to ideological consumption, wherein individuals purchase objects that match their beliefs.

From a neuroscientific perspective, a recent study at Yale revealed that when we think about objects we own, the same areas of the brain are activated as when we think about ourselves.[18] Our possessions are more than physical objects with a standard market value; they're part of our self-identity and therefore hold a value that can't be measured with money.

> *Our possessions are more than physical objects with a standard market value; they're part of our self-identity.*

If a stranger handed you a Montblanc pen and a scrappy BIC pen with a chewed-up lid, you'd have a pretty good idea which one was more valuable to them. If he then told you that he and his wife had signed their marriage certificate with that BIC, they've held on to it for 12 years, and they use it only once a year to write anniversary cards to each other, you'd probably change your mind. That's the power of meaning. It can totally transform the value of a product.

Storytelling for Nonstorytellers

Many people think that some products are just lucky enough to have a story and others aren't. But that's just not the case. There are a number of ways of adding a story to your product that will add meaning in the minds of your customers.

Heritage

If your product has been around for a while, it's bound to have an interesting story about its journey. You can associate it with the historical events that it has seen, the people who've relied on it, and the things that have come and gone in that time. Longevity is a great thing to be associated with; it adds a level of trust. If your product has been around for a long time, it must be good. Like the silver-haired wise man in 1970s kung fu films, we associate longevity with experience and wisdom. And that adds a level of credibility and trustworthiness.

If you're a gin drinker, you may have noticed a lot more bottles appearing behind the bar in recent years. One of those is Hendrick's, a small-batch gin created by a Scottish distiller. As a brand, it looks like it has been around since the days of Dickens, but the truth is quite different.

The gin was launched in 1999 as an alternative to the London dry gins that dominated the market. Traditionally, gin was the product of a single distillation, while Hendrick's is a blend of two separately distilled gins with the added essences of cucumber and rose.[19] There was nothing like this on the market, and it was being launched in direct competition to some very established and historic brands. This was an innovation.

However, instead of positioning this as something new, the distillers packaged the drink to make it look as if it had been around since the 1800s. The bottle has the design aesthetics of a Victorian apothecary, and the marketing materials look like they were created before the invention of the automobile. The fonts the makers use would be familiar to a 19th-century printer, the imagery tends to be etched rather than photographic, and the colors feel aged rather than vibrant. This creates an illusion of history with a spirit of Victorian quirkiness for added character. It informs everything the brand does, from the website to events.

But the company has a certain degree of permission to take this historic approach. William Grant & Sons, the whisky-maker that created the gin, has been around since 1886. And the two stills they use to distill the spirit date to 1948 and 1860.[20] The Victorian era was also one of great experimentation, which fits in beautifully with the product itself.

Since the product's launch, Hendrick's has taken care to protect this approach. The look and feel of its marketing has remained consistent. And its events continue to be influenced by quirky Victoriana. Its brand presence only gets stronger as it builds on the consistency of its previous work.

By leaning on a history that it never actually lived through, Hendrick's successfully created a product and brand that didn't just succeed in entering an established market, it kicked off a renaissance in gin-making. Scotland now produces over 70 percent of the gin consumed in the UK. And nearly 100 distilleries have opened in the country since 2010.[21]

Whisky's days may be numbered.

The Birth Story

Regardless of the age of your product, there's often a story of how it came to be. Maybe it was created to offer the market a better option. Maybe it was designed to appeal to a different market. Or maybe it was an accidental discovery. Whatever the reason, there's bound to be a story to tell—and hopefully, one that helps to reinforce the origins of your signature elements.

The Human Story

There's always a personal story behind a product. It could be the story of the designers, or the company's founders, or the people on the production line. Or the story might come from outside the company. It could be the story of the users, the salespeople, or product reviewers. In the best cases, it will involve the signature elements—telling the story of why they were developed or the difference they make. Human stories are more likely to create an emotional connection, so this can be a powerful route to take.

The Care that Goes into Creation

How is the product manufactured? How much human attention goes into it? Do people spend time crafting it or checking it or polishing it or personalizing it? People value the human touch. The myth that fine cigars are rolled on the thighs of virgins has persisted for this very reason. People don't want to know that their product is one of 8,000 spat out by a machine every minute and passionlessly rammed into a box by a minimum-wage worker in an industrial warehouse in Baltimore. Tell the story of the moments when people gave the product—or the signature elements—a bit of love. Even if the moment was simply at the planning stage when a designer pained over the ergonomics or the position of a button.

The Grand Mission

Increasingly, brands are talking about a higher purpose. Sometimes that's charitable, like TOMS Shoes. Sometimes it's a social movement, like Dove's Campaign for Real Beauty. And sometimes it's about disrupting an industry to make things better for the consumer, like many of the Virgin brands. These are purposes that are higher than just shifting units. And that helps passionate audiences get behind the company and what it produces.

The Hero's Journey

The classic Hero's Journey narrative has pivotal moments that change the direction of the action. The most dramatic of these is the ordeal that brings out the hero's courage and leads to a successful conclusion. There may be a moment of pain or loss or a sudden realization that led to your current product. Tell that tale and take people on the journey. The pain and triumph make the reward so much more meaningful.

The Story of How People Use Product

Think of the way you get ketchup out of a glass bottle. Do you hold it at a specific angle and tap it in a particular place? Do you hold it vertically and slap the bottom? Or do you just stick a knife right up in there to get the good stuff out? What about when you remove the gas nozzle from your car? Do you give it a little rattle to remove the last drops? When you're replacing paper in your printer at home, do you tap the stack of paper on your desk to align the sheets beforehand? These are all little rituals that people can identify with. So is there something special about the way people interact with your product? Do they open it in a particular way? Do they hold it in a specific way? If not, can you show them a unique way of doing it? Rituals become enjoyable experiences, and you may be able to use them to help people build a meaningful connection.

Just make sure your story is inspiring and relatable for your audience rather than a self-interested egocast. If the audience thinks you're interested only in yourself, you'll never make a deep connection.

Innovate the Benefits to Deliver Delight

Lots of brands see the sale of a product as the end of the transaction. That's their sole focus, and they don't care about the customer beyond the cash arriving in their bank account. That's a big mistake, and they're missing out on massive opportunities to resell, cross-sell, and turn their customers into powerful advocates. This is a big opportunity for any business trying to improve its Net Promoter Score (and who isn't these days?).

Smart companies understand that the most powerful piece of marketing you can create isn't a TV ad or a poster campaign or a fancy website—it's the product itself. If you create a product or service that makes people feel something, there's a good chance they'll talk about it. But that involves going way beyond the "delivering customer satisfaction" approach that most companies have. It means designing moments of delight for the user.

Innovate Your Product's Existing Benefits

Look at your product's existing benefits and ask yourself if you can make more of them. Can you draw attention to them more clearly? Can you add theater to them? Can you increase the amount of delight they deliver? You may already have valuable benefits you can make more of.

Think of the Nike Air technology. The air pocket was originally hidden inside the sole. When Nike innovated to make it visible in the first Air Max, it suddenly took off. Nike continued to innovate on the product until the air pocket extended beyond the heel to cover the entire length of the shoe. This has continually added drama to the product and kept it fresh.

This constant innovation has also made it harder for competitors to get a leg up on them. Reebok launched its Pump sneakers the year after the Air Max 1 in an attempt to claw back some market share. But they've never had the same success.

It's harder to hit a moving target.

Nike has turned its air pocket into an innovation platform. It changes with every product release. And the design opportunities are seemingly endless. This is the ideal situation for a signature element. You should be looking to evolve it and enhance the benefit with every iteration.

Create Pivotal Moments of Delight

Think back to the first time you actually got your hands on an iPhone. It probably belonged to one of your early-adopter friends who was showing you something that made you go,

"Wow." Delightful moments are shareable (not just in an online way) and turn your users into salespeople.

If you stop caring about customers as they start walking away from the checkout, you'll never benefit from powerful moments like these.

If you want your purchasers to feel something, you need to look at the pivotal moments in how they interact with your product.

But not all moments are equal. Out of thousands of interaction points, BMW identified 20 that matter most in terms of brand love and brand disappointment.[22] These were the ones that had the highest level of positive or negative emotion associated with them.

What might these be for your product? Consider product demonstration, the moment of purchase, delivery, unpacking, first interaction, continued interactions, and even trade-in and disposal. Great products continue to surprise and delight their users long after the credit-card transaction. Map out every interaction you can think of across the largest wall you can find in your office and highlight the ones that elicit the most feeling. Then choose one or two of these to develop into delightful iconic signature moments.

Own the Benefit

Great signature elements, whether they're product features or signature moments, provide benefits to the user. They make the user's life better in some way, even if just for a second.

Over time, iconic products and services become the standard-bearers for their category benefit. Companies that apply an Iconic Advantage® strategy are focused on relentlessly owning and innovating this benefit. And they'll keep doing that, even if it involves sacrificing or cannibalizing their existing businesses.

Millions of words have been written on the success of Amazon. It's an extraordinary company that's cornered the mass market and brought us innovation after innovation. (You may even have contributed to its profits by buying this book, and now you're reading it on one of its Kindles.) People usually point to the secret of its success as being its recommendation engine or mobile applications or reseller program. These have all played a part in the company's meteoric rise. But Amazon's real genius is far more fundamental than any of these things. It's 1-Click.

1-Click is so much more than a yellow button. It's the first part of Amazon's strategy to shorten the distance from "I want it" to "I have it." Shopping on the internet traditionally involved a virtual basket, a lengthy checkout process, and some confusing checkboxes that could potentially be adding you to a spammer's address book. Amazon's idea was to bypass all of that and improve the experience for its customers.

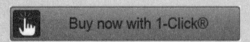

1-Click became the very soul of Amazon. The company recognized its power and innovated on the idea to extend it to every part of the Amazon universe. It's how you buy books on a Kindle. It's how you order with their mobile app. It's how you rent movies on Amazon TV. Amazon even created real buttons you can stick around your home to order products with one physical click. Same-day deliveries and physical "grab and go and pay later" stores naturally followed. And now with the company's planned purchase of Whole Foods, you'll be able to get your organic muesli without ever having to leave your yurt. The next development came with the increasingly popular Amazon Echo.[23] It allows people to order products with a simple voice command, giving Amazon a no-click service.

Amazon is now the world's biggest retailer. Part of its success is due to the 1-Click innovation and how the company has extended it to everything it does. That's helping it to own the larger benefit of "no patience required."

We won't even remember that Amazon was once a website.

Amazon's goal is to innovate and own the benefit of "no patience required," to the point of cannibalizing their existing business models. Don't be surprised if one day, with the integration of predictive analytics and IoT (the internet of things), items will show up before we even know we need them. And because of this, we won't even remember that Amazon was once a website.

Like the great river it's named after, Amazon seems to be an unstoppable force.

Amazon's 1-Click: Owning Instant Gratification

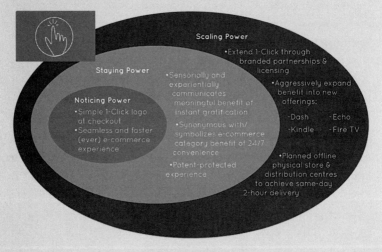

Scaling Power

•Extend 1-Click through branded partnerships & licensing

Staying Power

•Sensorially and experientially communicates meaningful benefit of instant gratification

•Aggressively expand benefit into new offerings:

-Dash -Echo
-Kindle -Fire TV

Noticing Power

•Simple 1-Click logo at checkout
•Seamless and faster (ever) e-commerce experience

•Synonymous with/ symbolizes e-commerce category benefit of 24/7 convenience

•Patent-protected experience

•Planned offline physical store & distribution centres to achieve same-day 2-hour delivery

Our idea of what makes up the **Iconic Advantage** of Amazon's 1-Click.

When we say "innovate the old," people sometimes interpret it as "don't do anything new." But that's not the message we want to give at all. It's all about focusing your innovation efforts in the right place.

When you own a category benefit, like Amazon does, you should continue to innovate on it. You should be looking to bring it to life in fresh ways. Occasionally that means treading on the toes of other areas of your business. Amazon regularly does this. Kindles eat into physical book sales. Amazon TV eats into DVD sales. The company is happy to cannibalize if it helps increase its differentiation. In the long term, it's the right thing to do for its business.

If you don't keep innovating on the category benefit, including innovating on the business model of your existing strengths, then someone else will. And that's how you lose your leadership position.

Reimagine Design to Build Excitement

This is about keeping your iconic product fresh and contemporary. If it doesn't change over time, it sends the message that you're not that interested in it, so maybe your audience shouldn't be either. That means it will slowly slip out of relevance and you'll lose your audience. How often you do a refresh depends on your industry. In some parts of the fashion industry, new product lines are launched every couple of weeks, while in the car industry, manufacturers may create seasonal limited editions, redesign a vehicle every three or four years, and launch a new car once a decade.

Refreshing your product is about more than just keeping it up to date. A 2006 study at University College London shows that novelty activates dopamine pathways in the brain, making us feel good.[24] Our brains crave dopamine. But over time the effects of a new stimulation wear off and less dopamine is produced. Products quite literally become less rewarding over time. Without frequent refreshes, even the most iconic product will gradually lose its wow and become just an everyday object.

Remember the feeling when you unboxed your latest smartphone? What feeling did that same phone give you a week later? Or a month later? Or six months later? The new-product dopamine buzz wears off quickly, because humans are adaptable.

If you don't keep innovating on the category benefit, including innovating on the business model of your existing strengths, then someone else will.

It's important to note that the effect you get from a new version of a product is not directly related to how different the updated version is. That's because humans crave both familiarity and newness. There's a balance to strike. It should be sufficiently fresh for people to notice but not so different that it feels alien.

Think of the iPhone. Each version looks pretty similar to the last. And each version of the operating system works in pretty much the same way. The best way to describe the process is refinement and evolution. There's very little unlearning to do when you trade in for the newer model. There are occasional "ooh" and "aah" developments, but the essence is always familiar.

The Right Amount of Newness

Your Iconic Brand Language will provide guidance on how to protect your familiarity while informing which elements are worth playing with. It will help you build on your past to develop your future. It will tell you what signature elements you can't mess with while showing you the opportunities for new expression.

There are two main areas that you can focus on to keep your products fresh and exciting.

Evolve the Product

Aim to draw attention even more strongly to your product's signature elements. Try to make them more noticeable, more powerful, and more appealing. This should be about further differentiating your product from the competition and building a stronger connection with your audience.

In addition, keep the styling fresh and contemporary. If your brand becomes unfashionable, you can seriously impact your desirability. Just make sure that your styling doesn't obscure your signature elements.

Allow your designers the license to explore the zeitgeist of today's fashion trends, pop culture, and social movements. Encourage them to infuse the freshness they uncover into the design aesthetics without sacrificing your signature elements.

Find the Right "Iconic Mash-up" Partners

One way to keep your brand fresh is to have an "iconic mash-up" strategy. Fashion and car brands are particularly good at this. The most successful ones manage to incorporate other brand properties without losing their own.

Nissan created a limited-edition Batman tie-in for the film *The Dark Knight Rises*, featuring matte black paint, like the Batmobile, and subtle bat insignias on the upholstery.[25] It succeeded in making the car look cooler than normal and attracted positive attention for the brand and product through the association.

In the world of fashion, Vans stands out for its flourishing partnership strategy. It augments its distinctive designs with special-edition brand partnerships. As of the time of this writing, it has partnerships with Pixar, Nintendo, and a number of high-profile designers. This is an integral part of its product strategy and generates significant profits.

If you can pull your eyes away from the hip haircuts of today's coolest kids and take a look at their feet, you're likely to see a pair of Vans on them. It's quite impressive how Vans has managed to stay fresh and relevant for 50 years. When it started out, it was known as the Van Doren Rubber Company and—rather unusually—it manufactured shoes and sold them directly to the public. On its very first morning, twelve customers purchased shoes that were made that very day and ready to be picked up in the afternoon.[26] These days Vans sells over a million pairs each year, many of which are the original design made on that first day of business. All because the shoes were quickly noticed and embraced by a niche market.

It just so happened that the unique waffle-shaped rubber sole was found to offer extra grip for skateboarding tricks. Very quickly the shoes were adopted by one of the most fashion-conscious communities. Now Vans just needs to make sure it held on to its fickle market. And that's something it has done amazingly well. It has proved that it has both **Noticing Power** and **Staying Power**.

Vans has continued to build a powerful relationship with its market by supporting street culture and creating the annual Vans Warped Tour concert series that's been bringing up-and-coming and established talent in a range of musical genres to the coolest audiences on earth for over 20 years. But its real stroke of genius is in the collaborations that help it stay on top of the latest movements in youth culture. If you look back through the company's history, you can find vintage shoes featuring designs based on timeless iconic franchises including the Beatles, the Simpsons, Star Wars, Hello Kitty, Major League Baseball, and Disney, as well as rare collections in collaboration with famous designers such as Kenzo, Marc Jacobs, and Takashi Murakami.[27]

Recently, the creative leadership team at Vans took the company's **Staying Power** to another level. Instead of just working with famous collaborators, the team tapped into the creativity of its community by updating the Vans Customs platform for people to create one-of-a-kind designs using Vans' iconic shoes as their canvas. This is an amazing way to make the brand even more relevant to its audience and give customers a deeper emotional connection with the Vans brand. Custom Vans shoes are so popular that they have become a canvas of choice for artists, who often sell their uniquely printed shoes on Etsy. Now that is **Scaling Power**. Naturally, this has resulted in a huge amount of valuable PR and social media noise.

It's this creative approach to being iconic that's helped Vans stay on top of a constantly shifting market. As long as there are skateboards and rock 'n' roll, it will have an audience waiting to buy its products.

Vans collaborated with the acclaimed Japanese contemporary artist Takashi Murakami.

Applying the Matrix

You may now be wondering how best to apply these approaches to an existing product or service. Obviously, you'll get the best results by addressing all four areas—but your area of expertise, budget, or resources may not allow for that. That's still OK. Addressing even a couple of quadrants can still have an impact if you do it well.

However, there's one quadrant you should look at before any of the others. And that's protecting your signature elements.

Once you've developed your signature elements (or agreed on your existing ones), you should create an Iconic Brand Language document to keep them safe. This will help to preserve your product's most valuable features and provide criteria to guide your decisions. Just as important, it will show you where you can play to keep your product fresh and exciting. We'll be looking at this in more detail later in the book.

How you focus on the other quadrants depends on what you want to achieve.

Building Relevance from Scratch
Iconic Relevance Matrix™

You can also use the matrix to guide you when you're developing a brand-new product. Again, we recommend that you undertake the steps in a specific order.

1. **Build strong Noticing Power** tied to clear points of difference that are embodied in signature elements.
2. **Protect your signature elements** by creating an Iconic Brand Language guide.

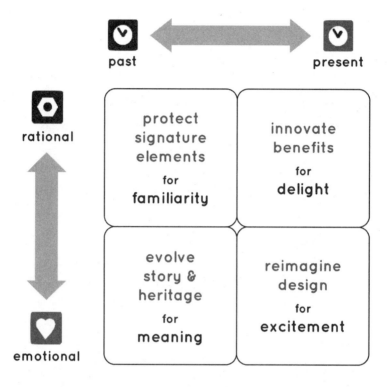

3. **Evolve your story and heritage** to connect more
   deeply with your audience.
4. **Innovate on the benefits** to give your audience more
   moments of delight.
5. **Reimagine the design** to keep your product fresh and
   relevant.

The Staying Power Toolkit

Working on your **Staying Power** takes a level of honesty that many people in business find uncomfortable. But if you're not brutally truthful, the marketplace will be brutal with your profits. Here are some exercises to help you build a connection with your audience over time.

Develop a Consistent Story

Hold one-on-one interviews with key employees throughout your company to discover the stories they associate with the product—like who developed it, why it was developed, how long it has been around, and what makes it different from the competition. Document the stories and highlight the most frequently told ones. Then repeat the process with your customers. Map these to see how consistent they are. Then work on creating the consistent story you want everyone to understand.

Create Delight

Map out each occasion that your audience comes into contact with your product and company—everything from marketing campaigns, the website, the store, the packaging, everyday use, and customer service, right up to trade-in and replacement. Identify the pivotal moments in the journey, the ones that elicit the highest emotions with your customers. Then take one or two (but no more) of those moments and come up with ideas for how you can innovate on them to make your audience feel delight—create something that exceeds expectations. It can be as small as a cute response to clicking a button on the website to building drama with the unboxing experience. Every touch-point is an opportunity—especially customer complaints.

Time-Travel

Imagine you can travel five years, 10 years, and 25 years into the future. Which world-changing developments could have an impact on your product? What about 3-D printing? Self-driving vehicles? Longer lifespans? Develop a fictional version of your product for each of the scenarios. How many of these future developments would work for you now? If you need to evolve into these product versions, what would each development look like? And most important, what would remain the same even 25 years from now?

Brand Ambassador Test

No matter how well you know your business, you are not a typical member of your audience. Invite some of your most loyal customers in and ask them how they feel about your product extensions. Would they feel the same pride about owning them? Would they be happy to be seen with them? Would they get the same joy out of using them? If not, why not? Use this to inform your design process.

CHAPTER 5

Building Scaling Power

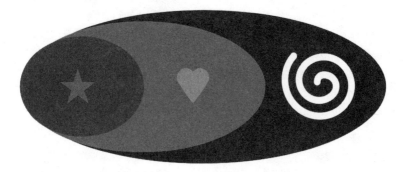

In 2013, Disney's princess movie *Frozen* played in pretty much every major movie theater around the world. It earned an astounding $1.3 billion in the box office and went on to become one of the highest-selling DVDs of all time.[1] But even if you've never seen the film, you're unlikely to have escaped the reach of its merchandising. It was shouting at you from the backpacks of a million eight-year-olds, selling you fast food at every mall, and serenading you from loudspeakers in ice rinks, elevators, and supermarkets all over the world.

At one point Walmart stocked over 700 items related to the movie, while Toys "R" Us held over 300 items.[2] So it probably comes as no surprise that the film's merchandising has now generated significantly more revenue than the film itself.

This isn't just entertainment; this is a business success that most Fortune 500 companies can only dream about.

However, this also was no accident. The real magic is how successful Disney was in cementing universal recognition for the franchise's signature elements. It is an expert at developing Iconic Advantage® that stands the test of time. Four years after *Frozen*'s release, the film's popularity doesn't appear to have waned.

It all started long before the red carpet was rolled out and the celebrities streamed into the premiere. Disney had identified the key signature elements of the film. Those went way beyond the lead characters of Elsa, Anna, Olaf, Kristoff, and Sven. The signature elements also included the songs, the audio, the best quotes, the props, the backdrops, the typography, the colors, and the logo. Disney created a repository of assets and defined how to use each element correctly.[3] This formed a powerful Iconic Brand Language™ that became the foundation for *Frozen*'s Noticing Power and Staying Power.

Then the merchandising and partnerships kicked in to take the universal recognition to whole new levels. Long before the film was released, warehouses were filled with millions of *Frozen* products featuring the signature iconic elements. In many ways, the film was the advertising for the merchandise lining the shelves of your local toy store. It was the vehicle that created universal recognition for the franchise's iconic elements. This is Scaling Power turned up to 11.

Disney currently has about fifty franchises, including all the Marvel characters, Pixar films, and *Star Wars*. In 2014, eleven of these franchises generated over $1 billion each. Now, more often than not, Disney makes more from merchandising and licensed products than it does from movies.[4]

By focusing the business on expanding and reinforcing each of its iconic franchises, Disney has truly got its magic back. A chart of its income over the past few years shows an increase so steep, it would give even the most ambitious accountant vertigo.

> Disney has come to understand the importance of protecting its iconic brands. It has learned how to scale their value without diluting them. And that's the real wizardry at the heart of the magic kingdom.

What Is Scaling Power?

So far, we've looked at Noticing Power, to create distinctive and memorable signature elements, and Staying Power, to build relevance through deeper connections with your audience over time. Now we'll look at building universal recognition so that you become the standard-bearer in the category you compete in.

We call this Scaling Power. This is how you become recognized as an industry icon for what makes you distinctive.

It includes advertising, promotion, distribution, product extensions, partnerships, and everything else that raises awareness, opens up new markets, and generates more opportunities to create universal recognition. Usually, this is approached with a "spray and pray" mentality. But as we'll see, that's not necessarily the appropriate strategy.

Get the Order Right

Don't be fooled. You can't just jump to this step and make your product a success—at least not in any sustainable way.

The three steps of creating Iconic Advantage are in a specific order for good reasons. You have to do them in the right sequence for this to work. Skipping to Scaling Power means that you'll be putting your product in front of your audience before it's ready. And that's a surefire way of damaging your credibility and making it harder to build any Iconic Advantage in the future.

In our experience, this is where a lot of companies get it wrong. When we ask them to list their current projects and priorities, most of them seem to concentrate on activities that fit into Scaling Power. To be fair, it's pretty understandable, because it seems like the most businesslike approach. Being that it's the last step in the process, it's easy to mistake it as the cause of any success. And because many of the activities involve significant budgets, they tend to come with metrics that take full credit for any revenue generated.

Executives are under increasing pressure to make more money each quarter, and they've historically tackled that in a number of tried-and-tested ways: through marketing and promotion, by rapidly expanding distribution, or by launching a flurry of new product extensions. However, these are expensive approaches that on their own don't do anything to build a deeper relationship with the customer or make the product more desirable.

In many markets, this can lead to your brand becoming ubiquitous and then irrelevant. It's the boom-then-bust approach. You may ship more product in the short term, but you're doing very little to create long-term sustainable growth.

This is a particular risk for fashion brands. In recent years, Coach and Michael Kors have fallen into this trap. By becoming ubiquitous and creating products for everyone at every price point, they've saturated the market. That may seem good for a while, but it has left them with a weaker foundation from which to grow their diluted brands.

Robin Lewis explains the risk Michael Kors faces in an article for *Forbes*: "All of a sudden, in a nano-split second, the largely young and trend-fickle consumer base wakes up and realizes the brand is slapped on everything and is being worn by everybody, everywhere. And, crash! Wonderful becomes awful. The brand stands for nothing for anybody–everywhere."[5]

These brands have diminished their iconic power by spreading themselves too broadly and becoming less meaningful.

As an illustration of how Coach lost its place, in 2014 it was reported that around 70 percent of its income was coming from outlet stores.[6] It's nearly impossible to build any meaningful relevance in that situation.

How Big Is Your Universe?

The word "universal" might sound as if we're talking about a mass-market, tell-everyone-and-their-mother approach. But it's actually about being far more focused than that. It's a waste of time trying to make your product iconic for people who probably won't buy it. So when we talk about universal recognition, we're talking about deeply and narrowly exposing it to your specific audience. No one else matters.

Like most people, you've probably never bought a container of Mr. Zog's Sex Wax (it's not what you're thinking). Chances are, this is the first time you've ever heard of it, unless you're a surfer. That's because Mr. Zog's Sex Wax is one of the most iconic brands of surfboard wax (which is the gunk people put on their boards for extra grip). It doesn't matter that you're unaware of it. It's universally recognized by the audience of surfers who matter.

And this is why you really need to understand your audience. Only if you know their needs, their passions, where they shop, and how they consume media can you truly connect with them.

If your audience is niche, focus on that. If your audience needs you only at specific moments, concentrate on those. If your audience is the mass market, go for mass media and mass distribution. If you haven't got your audience properly defined, you're simply wasting your time.

When we talk about universal recognition, we're talking about deeply and narrowly exposing it to your specific audience. No one else matters.

At the beginning of the book, we mentioned Roger Martin's strategic framework, and this is an important part of your "where to play" strategy.[7] You need to make sure you have a tight and well-defined **Scaling Power** plan that focuses on your audience. It's more important to be relevant to a small group of believers than it is to be nothing to everyone. Go narrow and deep rather than broad and wide. Or, to use an analogy: when spreading peanut butter, it's best if you make it extra thick and chunky. That's the way to build your **Iconic Advantage** and save money at the same time.

Iconic Advantage Framework™

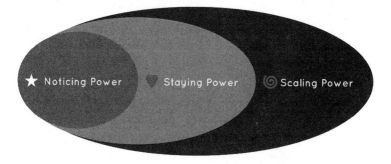

★ Noticing Power ♥ Staying Power ◎ Scaling Power

Why Is Universal Recognition Important?

Remember that iconic products become symbols that represent the people who buy them. For the products to earn that power, their distinctive signatures need to be familiar to their audience; those signature elements should be known and admired by everyone in the group. If your peers don't know about a product and what it stands for, it's not iconic.

Imagine a teenaged boy who is into fashion. He has to choose between two identically priced T-shirts with a prominent brand logo on the front. One reads "Under Armour," and the other reads "Red Cougar." Everyone has heard of one of these brands, and no one has heard of the other. Which one do you think he'd choose? Probably not the Red Cougar one (because we just made it up). The inherent value of the Under Armour shirt is that people recognize it, believe it to be high quality, and know that it's expensive.

When spreading peanut butter, it's best if you make it extra thick and chunky.

The more recognizable a product is to its audience, the more desirable it becomes—if it's recognized as being good, of course. It becomes a way of helping individuals identify more strongly with their group, whether they're middle-aged golfers, lovers of high-end fashion, indie-rock skateboarders, or members of whatever other niche you choose.

And once you've built your Iconic Advantage for a product line, you can use it to build anticipation of forthcoming releases. You already have your audience's trust and interest, and the rosy glow of your Iconic Advantage will be naturally transferred onto the new product. That's what makes people line up around the block and camp out overnight to be the first to claim the power your product offers them.

The Science of Recognition

There is a much-documented psychological principle called the mere exposure effect.[8] Quite simply, it describes how people tend to be more predisposed to a stimulus simply if they've been repeatedly exposed to it. And it follows that the more an audience comes into contact with a product or brand, the more it works its way into their minds.

Much of this happens subconsciously, which makes it even more powerful. As Daniel Kahneman made clear in his book *Thinking, Fast and Slow*, most decisions we make are fast, instinctive, and emotional. Iconic Advantage taps into this insight.[9]

The entire field of behavioral economics is based on the fact that people use heuristics—or mental shortcuts—when they make decisions. And that's very apparent in purchase decisions. One such example is the availability heuristic, which Wikipedia describes beautifully:

"The **availability heuristic** is a mental shortcut that relies on immediate examples that come to a given person's mind when evaluating a specific topic, concept, method, or decision. The availability heuristic operates on the notion that if something can be recalled, it must be important, or at least more important than alternative solutions which are not as readily recalled."[10]

As we've stated, products with Iconic Advantage are the ones that spring to mind when an audience thinks of their category. This is what's known in marketing land as unaided awareness. It's one of the best indicators of purchase consideration.

Many of our purchase decisions are heavily influenced by our peer groups. Studies show that we often buy products to communicate something to the other members of a group. The Prius effect mentioned earlier shows that people will spend a significant amount of money to show their social group how green they are.[11] But, of course, this kind of conspicuous consumption isn't limited to just environmentalists. It affects a wide swath of product choices at every price point.

If your product is favorably recognized by your audience, sales will follow. So when building Scaling Power, the goal is to focus on building favorable recognition. And this favorability is much easier when you have strong Noticing and Staying Powers to begin with.

Noticing Power is about selecting the best seeds. Staying Power is about planting them in the hearts of your audience members. Scaling Power waters them, adds the fertilizer, and gives them the sunlight they need to flourish. And together they give you a bumper harvest.

The Ingredients of Universal Recognition

When it comes to creating universal recognition, there are a few things you need to consider.

Repeated exposure to signature elements

As we've already seen, the more you see something, the more you trust it. And in many sectors, trust is an important qualifier for purchase. If you are consistent in the way you act and communicate (based on your Iconic Brand Language), you'll build confidence each time your audience is exposed to your signature elements. In short, you should be looking to get into your audience's eyeline as frequently as possible.

You can do that by buying ads in the media your audience consumes. Or you can do it in clever ways that aren't just about a big media spend. It's always better to outthink the competition than outspend them. One of the most powerful ways you can do that is to convert your loyal customers into your spokespeople. And maybe even your sales force.

Noticing Power is about selecting the best seeds. Staying Power is about planting them in the hearts of your audience members. Scaling Power waters them, adds the fertilizer, and gives them the sunlight they need to flourish. And together they give you a bumper harvest.

If you've ever bought an Apple product, you'll have discovered two logo stickers inside the packaging. Customers can put these signature logos on other objects to further associate themselves with the rosy brand glow of Apple.

Some people are pretty imaginative with their placement. And others just stick them around their desk and on everyday office paraphernalia. Regardless, it's still a lovely little bit of free advertising for the brand.

The kind of people who buy Apple products tend to hang around other people who are also likely to enjoy Apple products. Without any effort, the company has naturally targeted the right audience. The increased exposure to the logo reinforces the viewer's connection with the brand. And the stickers give each customer more opportunities to be seen as an Apple kind of person.

Quite a clever thing to do, I'm sure you'll agree. But—best of all—something that costs Apple next to nothing compared to the price of paid media.

The Apple symbol is a registered trademark of Apple Inc.

Consistent Messaging

Your messaging should be focused on drawing attention to your signature elements and their associated benefits. Don't waver from that. Noticing Power and Staying Power bake the marketing goodness into the product, and the role of your communications should be to make them as visible as possible. Don't let your signature elements become obscured by borrowed interest or promotional offers or stylistic flourishes. Please don't misunderstand us here—we're not saying you should eliminate creativity from your advertising and marketing. We're saying that the creative thinking should be focused on highlighting your signature elements and what makes them beneficial. After all, that's what will help you stand out, attract interest, and offer value to your target audience.

When Phil Knight started Nike, he didn't believe in advertising. So he ended up working with a new ad agency started by a couple of guys who were also pretty sick of advertising. They were Dan Wieden and David Kennedy. Nike was their first client.

Over the years, Wieden+Kennedy has become one of the world's most respected ad agencies. The campaigns it did for Nike helped to earn it that reputation. In an industry that often relies on borrowed interest and bombast, the agency focused on the products, their signature elements, and the benefits they offered. And the work it created succeeded in turning Phil Knight into a real believer. So much so that it transformed the way he viewed his own business. In his words, "We've come around to saying that Nike is a marketing-oriented company, and the product is our most important marketing tool."[12] That's the important insight: the product and its signature elements are the heroes of their marketing.

Take a look at some of the advertising for the Nike Air sneaker. The main signature element is the visible air cushion. And it's right there on display in every ad. Often the design intentionally draws attention to it even more strongly. Nothing obscures the marketing built into the product—it simply highlights it in a clean and stylish way.

It's a master class in how to showcase your **Noticing Power** while haloing the rest of the brand. And it's the secret to the success of both Nike and its legendary ad agency.

Noticing Power and Staying Power bake the marketing goodness into the product, and the role of your communications should be to make them as visible as possible.

Staying Relevant to Your Signature Elements

Your signature elements are what makes your products unique and desirable. But they may have a power beyond the products themselves—particularly if they involve an attitude or a way of interacting.

For example, as a company defined by 1-Click, Amazon would betray itself if it created an online promotion that required a long sign-up form to enter.

This is the equivalent of body language for businesses. What you do often conveys more meaningful information than what you say. Don't think that you can convince your audience just by using words.

The Three Vectors of Scaling Power

There are three areas you need to address if you want to develop effective Scaling Power. They are product/service extensions, marketing, and distribution. Each of them can have an immediate impact on your business. But you unlock their real power by using them to amplify your Noticing and Staying Powers. This creates a seamless Iconic Advantage strategy that will help to strengthen your iconic quality over time.

What you do often conveys more meaningful information than what you say.

1. Product/Service Extensions

The purpose of these is to expand the recognition of your Iconic Brand Language to more categories, thereby increasing your universal recognition. They can also create brand-new sales opportunities, either with your existing customers or by opening up entirely new markets. And there are a number of ways to approach them.

Tiering Up

This is about creating a premium version of your product. It can unlock new sales opportunities for you as well as give you the opportunity to add a product with a higher profit margin.

You need to start by understanding what people would be willing to pay extra for. That probably involves more than just smothering your existing product in Swarovski crystals and tripling the price. It's about understanding what would be valuable to existing and potential customers and upgrading your base product with these features. You're aiming to increase its desirability enough for people to be willing to pay a premium.

However, you need to do this without losing your signature elements. It shouldn't be an entirely different product—it should be a premium version of your existing product. And, again, this is where your Iconic Brand Language will guide you.

If possible, you should be looking to emphasize and celebrate your signature elements even more. They're the elements that offer value to your audience, after all. If you dilute them or eliminate them, you'll only confuse your audience and damage the Iconic Advantage you've already invested in.

There are a number of ways you can tier up your product. The fashion industry tends to approach it by using higher-quality materials and more attention to detail in the manufacturing. The tech industry tends to offer more functions, bandwidth, or memory. The service industry usually goes for added support. But it's all about finding what's right for your own products.

Apple has recently gone down the brand-collaboration route for its Apple Watch. The product remains the same, but the straps include Nike sports versions and superpremium Hermès leather options.

However you do it, you need to make sure you don't compromise your signature elements. This is your opportunity to add even more value and draw more attention to them.

There aren't many cars as iconic as the Porsche 911. Amazingly, half a century after it first hit the road, it's still just as desirable and sought-after.

To celebrate the 50th anniversary of the car, Porsche released a limited edition of 1,963 cars (that number is the year it was launched).[13] And it took the opportunity to pump up the car's iconic quality even more.

These specially designed cars have features that draw attention to their heritage and design. The 20-inch wheels nod to the legendary Fuchs wheels that have long been associated with the car. Classic design elements like the front air inlets, the fins of the engine compartment grille, and the panel between the rear lights are highlighted with chrome trim. The classic design features extend inside the car as well. The seats are upholstered in fabric similar to the Pepita tartan of the 1960s. And the instruments feature white pointer needles with silver caps, just like on the very first model to leave the factory.

This isn't just any old Porsche 911. This is the most iconic Porsche 911 around. Each of these elements has a story that goes with it. And that story is what makes this version of the car worth the higher price tag—that and the emotional thrill you get when you put your foot down on the gas pedal.

Tiering Down

This is about creating a more affordable version of your product with a lower price tag to access the mass market without diminishing your iconic equity. Fashion companies often do it. Even Apple did it with the iPhone SE. Companies create more accessible products to cater to people who aspire to own the brand but can't quite afford it.

There are a number of ways to approach this strategy. You can use less expensive components for nonessential elements. Or remove the nonessential elements altogether. You can cut down on the craftsmanship by automating more of the process. You can offer less support to your customers. Or you can simply rely on the reduced production costs that come with higher manufacturing volume. Again, however you approach it, you need to make sure you retain what it is that makes your product recognizable and loved. This is about taking your main signature elements as defined in your Iconic Brand Language and finding a way of delivering them for less.

The Fender Stratocaster is probably the most famous guitar in the world. For over 60 years, it has been the choice of guitar legends like Jimi Hendrix, Eric Clapton, and The Edge. But with the basic models starting at around $600, it's out of reach for most people. In the '70s, the only alternative for guitarists on a budget was to buy cheap knockoffs from overseas that looked similar to a Strat but didn't have the feel or the sound of a Strat. That all changed in 1982 when Fender created an affordable range of guitars under the brand name Squier.[14]

The Squier Stratocaster retained the key signature elements of the original. It had the same name, same shape, same pickup configuration, and same wood finishes. It was just made overseas from less expensive materials.

Squier guitars were comparable in price to the knockoffs—but they had the rosy glow of Fender around them. All the company's knowledge and talent went into creating these more affordable instruments, and the quality was head and shoulders above the hit-and-miss foreign fakes. The small "by Fender" statement underneath the Squier logo was just enough to make the products desirable to any fan of the brand.

Fender succeeded in tapping into a market it had previously excluded. And it did that by retaining all of the Stratocaster's signature elements.

Adjacent Categories

Whereas tiering up and tiering down are all about opening up new consumer segments, moving into adjacent categories is about meeting the additional needs of your existing, loyal customers. These can be new occasions to use your product or complementary uses. It's all about getting your existing customer base to buy more from you.

You do that by creating new relevant products that feature your existing iconic signature elements. This empowers your customers by giving them more opportunities to showcase the brand they love, further increasing your universal recognition. Importantly, you give your customers more reasons to stay engaged with a brand they love. And that can stop them from looking to other brands to fill their needs.

You need to start by defining what market or niche your product is serving. Then ask what else these individuals need. Look for the moments or occasions (before, during, and after) they use your product and list the other items they use around the same time.

A good example of this is Puma. It started as a sports shoe manufacturer. But through the years it has expanded it product line to cover everything else a sportsperson might need. It now offers clothing, bags, and a wide variety of sporting accessories. All of these additional products feature the Puma cougar icon and logo. And the items are designed to work together as a style. This creates a "look" that customers can buy into.

Likewise, KitchenAid started out with their iconic mixer, but over time it expanded to offer a range of accessories and tools to help with all aspects of food preparation. It now offers everything from spatulas to refrigerators, ice cream scoops to kettles. As well as featuring a shiny metal KitchenAid logo, these products feature the bold enamel colors and 1950s styling of the original mixers. The designs bring style into the kitchen and have succeeded in making the food mixers even more desirable.

To execute this strategy well, you should extend the signature elements in your Iconic Brand Language into your new products. They must be obviously part of the same family. And they should build on your beliefs and purpose.

Whatever additional products you create, they need to be built on your Noticing and Staying Powers. If they're not, don't do it.

Collaborations

This is about accessing an audience that another brand has already built a relationship with, and about leveraging the reputation and credibility of that brand. You do that by collaborating with them.

You start by identifying audiences that you'd like to gain access to—preferably audiences that you believe will be sympathetic to your product. Then you identify which brands they already identify with. Finally, you approach these brands to work on valuable collaborations.

In the fashion world, this is most commonly done by licensing a company's intellectual property. Step into H&M or Target and you'll see band logos, cool brands, and cartoon characters on the T-shirts. These products are borrowing desirability from these collaborations. But there are more ways to approach this.

Another route is to bring in celebrity designers to work with you on new products. Again, H&M is a master at this. The company's collaborations have included Karl Lagerfeld, Stella McCartney, Versace, and Jimmy Choo.[15] But this approach isn't restricted to just fashion. Since the 1980s, Beck's Brewery has commissioned artists to design limited-edition labels for its bottles. These design collaborators have included Jeff Koons, Tracey Emin, and Gilbert & George.[16] Range Rover has also used this strategy. In 2012 it launched a limited-edition "baby" Evoque designed by former Spice Girl Victoria Beckham.[17]

This approach is about mashing up two iconic properties and, in the process, improving the relevance of both and increasing the universal recognition of both.

Just make sure you don't dilute your own signature elements in the process, or do something that appears to be out of character or betrays your beliefs and purpose.

If you find the right way of collaborating, this can be a powerful way to get in front of a whole new audience and open up a fresh source of income.

2. Marketing

Getting your product in front of the right people is vital for building awareness—and therefore desirability. And that job falls squarely in the laps of the marketers. The cost and approach of your marketing depend on your category, audience, and ambitions. It may be that an expensive prime-time TV campaign is what you need. Or maybe a guerrilla campaign would be better for you. It's all dictated by your audience and what they find most persuasive.

The main goal of your marketing should be to raise awareness of your iconic **Signature Elements**. This will be guided by your Iconic Brand Language. It will influence every aspect of your marketing efforts, including the creative brief, the look, the feel, the messaging, and the chosen media.

Advertising

The lion's share of most companies' marketing spend goes toward advertising. And it's debatable how effective most advertising is. If you flick through a magazine or watch an evening's worth of ad breaks, you'll see a mix of boring messages and borrowed interest. Surprisingly few ads draw your attention to what makes a product distinctive.

If you have signature elements, you shouldn't ignore them. Everything you do should draw attention to them and dramatize their benefits. Iconic Advantage embeds the real marketing in the product, so the job of advertising campaigns is to showcase these crown jewels and to build a connection with the audience. Don't allow anything to dilute that. It's an expensive way to be ineffective.

It's Not Just About Paid Media

As the media landscape continues to fragment, there's often uncertainty about how to use the channels. But you're unlikely to go wrong if you use them to bring your Iconic Advantage to life. Show your signature elements, make your purpose clear, tell your story, and connect with your audience. That works for point of sale, direct marketing, digital banners, product demonstrations, PR stunts, social media, branded content, and whatever new channels develop in the coming months and years.

Iconic Advantage embeds the real marketing in the product, so the job of advertising campaigns is to showcase these crown jewels.

Red Bull is a master at this. It uses standard advertising channels, but it's better known for its events and stunts. Its signature purpose of enabling people to perform at their peak is brought to life in every activity the company does—most notably Felix Baumgartner's leap from the edge of space. Everything builds on Red Bull's story and helps it connect more deeply with its audience.

Your Iconic Advantage is the key to desirability, loyalty, and longevity. But you'll never unlock its real benefits if you don't place it front and center.

If you're familiar with ordering drinks in a bar, you'll know that Corona is served slightly differently than other beers. It comes with a wedge of lime tucked into the neck of the bottle. And if you don't get that fruity garnish, you may have a stern word with the bartender.

This serving suggestion is a great **signature element**. It adds a refreshing twist (literally and metaphorically) to the drink. It helps it stand out from the other amber liquids in the fridge. And it's a hugely sensory element, involving sight, smell, taste, and touch.

Corona understands this well. And it uses this signature element consistently in its marketing materials. You'll find it on the website, web banners, and point-of-sale materials as well as in TV ads, posters, and print ads. If a bottle of Corona doesn't have a wedge of lime in it, it's underdressed.

The marketing consistently draws attention to this signature element, helping to build a distinctive difference from the other beers on the market. And for Corona, an ice-cold lager with a hint of fresh lime is actually the taste of success.

3. Distribution

Part of creating universal recognition is making sure your product or service is available to your customers when they're looking for it. So choosing the right distribution channels is absolutely vital. If your product is hard to find at the crucial moment, all the work you've put in so far could be wasted.

However, you really need to go further than merely being available. The way that people find and buy what you're offering can also reinforce your Noticing and Staying Powers. This isn't just a rational business decision. This is a decision that needs to be informed by your Iconic Brand Language.

Retail Presence

There are two sides to this. The first, and most obvious, is that you need channels in place to get your product into the hands of the consumer. So, if your audience tends to shop in certain stores (online as well as physical), then you aim to get those stores to stock your product. There's nothing surprising about that. It's what most businesses already do.

A more sophisticated approach is looking at distribution as part of your strategy to communicate and reinforce your iconic signatures. The way your audience gets their hands on your product can be used to communicate something about it. And that can help you connect more deeply with your audience. It may be that niche distribution and limited availability work for you better than ubiquitous availability. Or maybe using outlets that are totally different from the competition's will help you differentiate yourself further and communicate something about what you stand for. It's not just about where your product is stocked but also why and how it's stocked there.

How did you buy your last car? If you're like the majority of people, you went to an out-of-town car dealership, where you had a look at the models, took one for a test drive, and haggled over the trade-in terms of your old car. That's the way it's done, right? Not if you're buying a Tesla. Its makers have created their own approach that's centered around retail stores in shopping malls and high streets. They hold no stock and don't employ your typical salespeople.

And they've got very good reasons for this.

Tesla is offering a new category of product, so its job is as much about education as it is about sales. All of the technology it's developing aims to reduce our environmental impact. So it makes sense that the way it showcases its cars reflects the same signature values.

The in-store experiences, like the cars themselves, are amazing feats of technology. Interactive screens allow you to explore content created by actual drivers, configure your own model, and see it in action. This creates a much more immersive experience in a smaller—and greener—sales environment.[18, 19]

And because the long-term goal is to replace the existing fossil-fuel car industry, Tesla has refused to employ experienced car sales-people. In its opinion, these individuals have such a history of selling the very cars that Tesla is trying to replace, that they wouldn't make believable advocates for the brand.[20] The company doesn't want any-thing to take away from its signature beliefs.

But over and above these practical differences, implementing a disruptively different retail strategy helps to express the Tesla car's key signature elements. Their retail approach is simply another chance to hammer home the message that Teslas are better for you and society because they leverage the power of modern technology. And the store experience does that beautifully.

Geographic Focus

The internet has removed most of the international market barriers that used to exist in predigital days. So it's increasingly important to make sure your product transcends borders and works across cultures if you want to tap into a global audience.

Some products work across all markets without any adaptation. If that's the case, bingo!

However, in many cases, your markets will have differences in taste and culture that need to be addressed. In that case, you need to consider two contradictory factors: one is consistency and the other is relevance. If you've created a robust Iconic Brand Language strategy, you'll find it easier to create the right balance.

You'll have defined the signature elements that can't be tampered with and outlined the areas where you can play. This helps you keep the important elements of your product or experience intact while adapting it for the specifics of each local market. Just make sure you don't allow dramatic inconsistencies, because people travel and use the internet. They'll quickly highlight anything they find abroad that contradicts their understanding of who you are and what you do. And that can put a dent in all the hard work you've put into building your Iconic Advantage.

Wherever you travel abroad, you'll usually find a McDonald's. You can't fail to recognize it. It looks exactly like the McDonald's you're used to at home and has a pretty similar menu.

But in many places, you'll discover some different menu items and slight variations on the usual recipes.

For example, in India, you can order dishes like the McPaneer Royale—with cheese instead of meat—and the Veg Pizza McPuff—which would not sell at all well in the US. Because India is a more vegetarian country, the menu has been adapted to reflect that—but everything is still done very much in a McDonald's way.

You still find the "Mc" beginning many of the dishes' names. You still get your fizzy sodas in a paper cup with a lid and a straw. And you can still top it all off with a McFlurry. You've got the full McDonald's experience—just a slightly localized version of it.

Tools to Test for Scaling Power
The Advertising Test

Show your advertising to ten people and ask them to tell you what stands out and what they think about it. If they don't consistently identify your product's signature elements, your advertising isn't doing its job. You should probably do this on a small, informal scale before you spend money on focus groups. Your informal test will make sure that the focus of the advertising is right; the focus group will make sure the ads communicate it in the right way.

Extension or Irrelevance

When you're developing product extensions, you want to make sure they build on your **Iconic Advantage**. Show the product extension to 10 people and ask them what product it reminds them of. If they don't identify your original, core product, you may need to redesign it to make the signature elements clearer.

Shelf Test

Place all your product extensions on a shelf to see how they look together. Ask other people what they think. If you want to really test things, place a dummy product in the mix that deliberately doesn't have the signature elements and ask people to pick the odd one out. If they can't consistently pick the ringer, you may need to emphasize the signature elements more strongly.

CHAPTER 6

Laying the Foundation
for Iconic Advantage®

The next time you're out and about, cast your eyes down to see what people are wearing on their feet. Unless you're in a financial district during office hours, you're bound to see a few pairs of Converse sneakers. You'll see them in a variety of colors, with different numbers of lace holes, and in special-edition designs. But regardless of these variations, they are still unmistakable. They are one of the most iconic shoe designs of the past hundred years.

These shoes have barely changed since they were first launched. The materials they're currently made of may be subtly different from the materials that were used when the shoes rolled off the production line 50 years ago, but the design has hardly been touched. Kurt Cobain and President Kennedy had a pretty similar view of their feet when they wore their Chucks. And that's pretty special.[1]

That's why it was surprising when Converse decided to revisit the design of its classic All-Star sneakers in 2014—especially when one of the company mantras is "Don't fuck with the Chuck."

The designers had quite a challenge on their hands. Converse had some pretty strong **Iconic Brand Language™** that was well understood by their audience. They were being asked to refresh the design of the shoe, but if they messed with the look too much, they were likely to alienate their loyal audience. So they set to work on finding a functional design element that made sense to their customer base of skateboarders, musicians, and all the other people who held the Chuck close to their heart.

They got their X-ACTO knives out and performed some thorough autopsies of the product line at the time. They defined the signature elements that were immovable and the design features that were flexible. They then started playing with the constituent parts to see what would happen. It took nearly a year to get to what they were looking for.

They didn't change the shape of the shoe in any way. That was sacrosanct. However, they added a design element that had an additional benefit to the lifestyle of Chuck wearers. They included reflective print in the design of the Chuck II to make them easier to see in low light. This would be a bonus feature for the skateboarders, cyclists, and general pedestrians who wear their shoes at nighttime.[2]

While they were at it, they added some extra comfort padding on the inside of the shoe, where it would be felt but not seen. And they made minor cosmetic alterations that didn't affect the overall look of the shoe.

They successfully updated the traditional Converse sneaker and added a significant functional design element. But most important, they didn't fuck with the Chuck.

Many brands are going through an identity crisis. They just don't know who they are or what they're about. Which makes it nearly impossible for anyone to feel a connection with them.

These businesses don't understand that real consistency is about being guided by your beliefs in every situation. If you have strong "whys" at your core, they will keep you steady and relevant as the world changes. This strong identity will allow you to act appropriately in any situation. It will allow your brand to become timeless by balancing the familiarity of the old with the excitement of the new.

Self-awareness comes from an understanding of signature elements, and of your brand's purpose, values, and points of difference. This knowledge will give you the confidence to make important decisions.

You can see the brands that don't have this knowledge. Instead of figuring themselves out, many of them rely on retreading the same tried and tested paths. They'd rather stick with the same old product design, just to be safe. They adhere to overly rigid guidelines that focus on what to do rather than why to do it. They mistakenly believe consistency is about repeating the same actions again and again regardless of the context or the shifting environment.

This kind of inflexibility in a changing world is a guaranteed route to failure.

But some brands react in the exact opposite way. Out of insecurity, they keep nothing consistent as they flail around trying to become relevant to their audience. Again, this comes from a lack of understanding of who they are and why they exist.

You can see them radically reinventing themselves in a desperate attempt to regain their former success. They're like 40-something men who suddenly go out and get a tattoo or a motorbike or a sports car. It seems out of character and shallow. It betrays a lack of confidence in who they were and who they've become. It's not guided by a deeper purpose.

Become timeless by balancing the familiarity of the old with the excitement of the new.

A very visible example of this is a comparison of the Coca-Cola and Pepsi logos through the years:

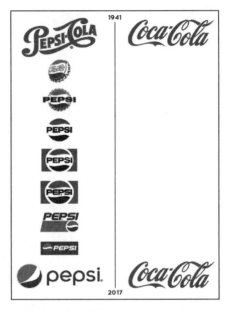

The evolution of the Pepsi and Coca-Cola logos.[3]

It's obvious which one of these brands appears to be confident in itself. Just look at the dramatic changes Pepsi has made to its logo through the years, the most drastic change being the current iteration.

The press releases refer to these new looks as "brand refreshes" and go on to tell people that they're doing it to "bring humanity back," to "make the logo more dynamic," and to create a look that's "more alive." These are actual quotes that a Pepsi vice president said to *Ad Age* about the 2008 refresh.[4] The dramatic new look came shortly after a new CEO took the reins with a very different strategy. It may have felt right for the company, but to the consumer, it felt like a desperate lover saying, "Please don't go; I can change!"

A quick straw poll of some of our (admittedly, not-so-young) friends and colleagues resulted in less than half of them knowing which logo was the latest iteration. At the time

You're not just changing the identity of your product; you're also affecting the self-image of every single member of your audience.

of this writing, the logo has been around for nearly ten years, and it's still not fully understood by the public.

Brands don't seem to understand how their inconsistent behavior feels for their audience.

This "quantum leap" approach (again a term Pepsi used to describe their refresh) is asking their existing audience to disconnect from an attachment they've already built up and automatically transfer their feelings to something that looks, feels, and acts entirely differently. It's just like when a new actor steps into James Bond's shoes; the audience holds back judgment before deciding whether that person actually embodies the character in a way they approve.

Of course, changes are necessary. You'll become irrelevant if you remain utterly static. You should aim to protect your signature elements while regularly infusing the brand with excitement and delight through design refreshes and innovation. The goal is to gradually evolve the brand without departing from its core character. When brands change their identities in seismic jolts, it's disconcerting for their audiences. This is especially true if you're an iconic brand that they identify with—because that means you're not just changing the identity of your product; you're also affecting the self-image of every single member of your audience.

It's important for products and brands to be consistent. But consistency comes only from knowing who you are. In the business world, that doesn't happen by accident. It's something that takes conscious effort and strong leadership. And you need to make sure that everyone is aware of it.

That's why you need to define your Iconic Brand Language. And that's what the next two chapters are all about.

Don't Try to Be Everything to Everyone

Having an identity and a belief isn't just about the way you look. It affects everything you do. And sadly, most companies show that their sole motivation is to make as much money as possible from as many people as possible.

The more broadly you define your audience, the harder it is to mean something to them. You end up being bland and superficial in an attempt to relate to the masses. And that's never going to build a meaningful connection.

When you have a well-defined audience with a common passion or interest, it's easier to be relevant to them. And that's vital if you want to build Iconic Advantage®.

That doesn't mean your audience can't be big. It just means it has to be well-defined. And successfully connecting with people involves your understanding your brand just as well.

Sowing the Seeds of Confusion

Even if the people at the top of the company have a vivid understanding of their brands and products, that knowledge doesn't naturally percolate through the ranks. Without defined guardrails and clear direction, companies naturally lose focus. Every employee comes with a different perspective, different motivations, and a different skill set. Left to their own devices, they'll interpret and express the spirit of a product in varying and often contradictory ways.

Some will be serious, while others will be irreverent. Some will want the logo to be big and prominent, while others will want it to be subtle and understated. Some will see the value of building an emotional connection, while others can't see beyond rational messaging.

You can spot it in many companies. They end up with too many logo treatments. They suffer from inconsistent product design. They don't have a consistent story. They don't appear to have a core belief. And even if they do have all of these things, they don't tend to be consistent enough in how they express them.

This behavior makes it almost impossible for their customers to connect with them. If a human acted as unpredictably, you'd probably do your best to avoid them.

These companies are inadvertently sabotaging their own success. They're preventing people from getting passionate about their products.

And it's only getting worse as communication, marketing, and sales channels continue to multiply.

Why Is this Issue So Common?

Most companies, especially the big ones, are overly focused on immediate revenue. They're running from quarter to quarter, trying to make more profit than they did in the same period last year, like an athlete constantly trying to achieve a new personal best. There's nothing wrong with financial ambitions. But when marginal quarterly gains become your prime focus, you tend to miss out on the important stuff that can really transform your business.

As a result of this approach, many businesses haven't even considered defining their iconic elements. This is partly because they don't see its leading to direct revenue. If they're having success, they take it for granted without ever really understanding why people are currently buying from them. To make it worse, many companies try to turn a successful franchise into as many new revenue streams as possible without understanding why it was successful in the first place. That puts them in a dangerous position and makes business decisions risky.

Fortunately, discovering and protecting your signature elements isn't complicated. Doing that takes just a little bit of effort.

We're going to show you how by explaining how to create an Iconic Brand Language document that is grounded in your brand's DNA.

Let Us Introduce You to *You*

It's always good to begin at the beginning. So let's start with your DNA. This is about getting to know what it is that drives your business, your employees, and your audience. You may already have done this. If that's the case, well done. If not, let's get going.

This is our Iconic Brand Pyramid™ and it's the foundation for building Iconic Advantage. You may have seen something similar (and probably a bit more complicated) before. For our purposes, you need to just answer four different but related questions. But you won't be giving quick-response, top-of-your-head answers. Doing this well requires a fair amount of chin-scratching and soul-searching. And you should get a number of people across your organization to do it as well. The broader the input you have on this, the better.

Let's start at the base and work our way up.

Iconic Brand Pyramid™

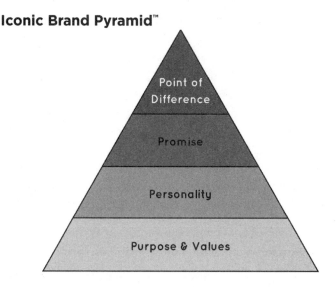

Purpose and Values

Purpose is what motivates your employees. It's why they wake up in the morning. It's what makes them look forward to coming into work.

Values are what guide your employees' behavior. They're the tenets and principles that direct their actions. They're what they draw on when faced with a dilemma.

Together, your purpose and values define how aligned your workforce is and how effectively each employee operates. The best companies know what these are and bake them into their culture.

You can't be cynical about this. It needs to come from the top. It's something the leadership should live by as much as the people on the frontline. It needs to permeate the entire organization.

If you struggle to come up with a purpose that's deeper than "because they get paid," you have a problem. In that case, you need to work out what it would be that would drive the workforce, and find a way of making that part of your culture and processes.

Your values are likely to be things like honesty, integrity, passion and putting the customer first. They're the moral standards that embody the purpose and mission of the company. They're the principles that guide your employees' behavior when no one else is looking. They are what helps them navigate ethical dilemmas and make the best decisions.

A good example of a response for this section would be:

"To connect people to what's important in their lives through friendly, reliable, and low-cost air travel."

This represents the purpose and values of Southwest Airlines. And you can see how it guides their behavior. The airline refuses to charge people for checked bags. That may not sound like the wisest of business decisions, but it's the right choice for their purpose. And it gives them a differentiation that they've highlighted in their advertising campaigns.

Personality

This is about the character of your organization. It's what drives you to do what you do as a business.

One of the best ways of looking at this is described in the book *The Hero and the Outlaw*, by Margaret Mark.[5] She describes four primary motivations:

- ▶ Independence and fulfillment
- ▶ Mastery and risk

- ▶ Belonging and enjoyment
- ▶ Stability and control

And below them, she offers 12 different archetypes that can help you define an organization's personality: innocent, explorer, sage, hero, outlaw, magician, regular guy/gal, lover, jester, ruler, creator, and caregiver. These are described in detail in Mark's book.

It may be an interesting exercise to get your colleagues from across the organization to choose the one or two archetypes they think best describe the business. The more aligned you are collectively, the more effectively you're working together.

If you're a hard-nosed, rational business-thinker, you may be questioning why defining your organization's personality is important at all.

The blunt truth is that people can't connect with a cold, unfeeling corporation. And connecting with an audience on an emotional level is a vital part of achieving Iconic Advantage.

To do that, you need to find an affinity with one or two of the archetypes described in *The Hero and the Outlaw*. Don't choose any more than that, or you'll come across as a schizophrenic brand that will struggle to gain anyone's trust.

Then you need to stick with it. Your audience will feel disconcerted if you suddenly change your personality or behave in a way that's inconsistent with their understanding of you.

Let's look at an example of a business that understands its personality.

If there's one hotel chain that really caters to the hipster crowd, it's the W. And it does so beautifully. This is what its website says:

"Guests are invited into surprising, sensory environments where amplified entertainment, vibrant lounges, modern guestrooms and innovative cocktails and cuisine create more than just a hotel experience, but a luxury lifestyle destination."

It's a hotel chain, but it defines itself as more than that. It's about sensuality, energy, and experience.

Having such a well-defined personality has allowed Starwood, the parent company of the W, to easily scale this brand into other related properties, such as the more affordable Aloft hotel chain and the XYZ bars in its hotels. This is a hotel to check out.

Promise

This is what the brand is committed to delivering to its users. It's your promise to them. This is the part of your DNA that the public actually gets to see or—even better—experience in a remarkable way.

It's informed by the other two levels you've already completed. And a good formula for it would be:

"Your brand promises...to do something remarkable."

This statement shouldn't be mundane or obvious. It shouldn't be based on rational features. It shouldn't just be a category promise. Instead, it should be something with an emotional benefit. Something that feels a bit special and different.

A good example of this is FedEx.

Back in the days when American kids used to walk five miles barefoot to school every day, the only express shipping available was offered by the post office. A delivery would often take a week to arrive. Of course, if you wanted international express, it involved waiting in line to make sure your package arrived within a couple of weeks.

However, that all changed in 1971. A new service company, Federal Express, was born, and the industry was never the same.

Not only did the company, now called FedEx, provide overnight delivery of packages, but it redefined the idea of customer-centric services.

The company's Purple Promise states:

"I will make every FedEx experience outstanding."

And this shows up in various ways, all of which are focused on giving the customer the best experience. It led to FedEx's creating the world's first global tracking network for shipments. It was the first company to let customers electronically request pickups and the first to allow customers to track their packages on the internet.

FedEx shows that a good promise can be a driving force for innovation.

Point of Difference

We've made it to the top. This is about the key skill, capability, feature, experience, benefit, or whatever else gives you an advantage over the competition. It's about what makes you stand out and become the preferred choice.

In short, it's what will help you win, not just compete.

If you can answer this right away, you're one of the lucky few. Many businesses end up playing in the commoditized end of the pool, where everyone offers pretty much the same thing. The fact that you've read this far indicates that you're not content to play in the dull and crowded shallow end.

You may have gathered by now that it's your point of difference that creates your Noticing Power. It's what makes your product distinctive in the marketplace. You should use the point of difference as your inspiration for developing distinctive iconic signature elements. If you succeed at this, you'll give yourself an advantage in the marketplace that leads to a larger share of attention from a more passionate group of customers.

We recommend that you don't have too many signature elements. Go for small and focused. One or two great signature elements are better than a dozen competing ones.

One of the most iconic silhouettes in the automotive industry is the Porsche 911. One of the things that influences the car's shape is what many automotive engineers would consider a major design flaw—the engine is placed at the very back of the car.

However, this has always been the 911's key point of difference.

Engineers will argue—correctly—that having the engine at the very back creates an imbalance of weight, which leads to a slingshot effect when making hard turns. Having your back wheels slip out from under you is a very disconcerting experience.

Yet, it's this very design flaw that makes the 911 special. You see, having the weight on the back wheels results in greater torque and better acceleration, and it requires the drivers brave enough to race the car to be fully alert at all times.

Importantly, the Porsche's designers have constantly innovated to create new signature technologies to compensate for this slingshot effect. Some of these are Porsche Stability Management (which keeps the car perfectly flat during a turn), Porsche Torque Vectoring (which selectively applies the brakes to the inside wheels for tighter turns), and Porsche Traction Management (which allows all four wheels to turn independently).[6]

This relentless focus on the Porsche's key point of difference is what makes it iconic.

Once you've completed this pyramid to your satisfaction, you'll have a better understanding of what lies at the core of your business. And that's a vital understanding to have when you start working on your Iconic Brand Language document.

Outward Success Starts with Inner Understanding

It's easy to see the brands that get it. They manage to remain timeless while also staying fresh and relevant. They develop familiarity and consistency while also extending and growing their business.

This isn't just about manufacturers. Service companies also need to remain consistent. And American Express has done this brilliantly.

An Amex card isn't the company's product. It's an embodiment of its product—which gets you better perks when you buy things. Putting an Amex card on the table says something about you. It says you're smarter and more professional than the person who puts their bank debit card on the table. And if you've got a Platinum Amex, it also says you're rich and well-traveled.

Amex's offerings have been utterly consistent over the years. The company has always offered better service, more perks, and bigger rewards than other payment cards. Whatever it does, that's at its core. It's central to the promise its customers believe in. And it forms a strong point of difference in the marketplace. If Amex changed these signature elements, it would be a betrayal of the customers who have remained loyal through the decades. And, of course, that would be devastating to the business.

Don't Skip Your DNA

We know you're probably excited to get on with building your Iconic Advantage, but it's not a good idea to start this until you understand your DNA.

In summary, you want your signature elements to be relevant, rather than just different for the sake of it. This relevance should be consistent with your brand DNA. And—even better—your signature elements should embody your key point of difference. Your point of difference should also reinforce your promise to your customers. And your promise should be consistent with your brand personality, purpose, and values.

Having all of these things in place will ensure that everything you do is aligned.

Your DNA is the foundation that your Iconic Advantage strategy stands on. Make sure it's solid.

CHAPTER 7

Capturing Your Iconic Brand Language™

Seven years before Ray Kroc opened his first McDonald's restaurant, Harry and Esther Snyder, a pair of quiet newlyweds, opened In-N-Out Burger in Baldwin Park, California. They had both served in World War II and were keen to start a small business that would help support a family. Their stand was tiny—barely 10 feet square—but their focus on authentic local quality was big.[1]

Harry's approach to ingredients was more like a five-star restaurant's than a fast-food joint's. As any good chef will tell you, the quality of a dish is defined by the quality of the ingredients. So every morning before dawn, Harry would do his rounds of the markets to pick up the freshest ingredients for the day ahead. Then he'd cook them with care in his cramped little kitchen, in full view of his waiting customers. He wasn't trying to be all things to all people—his menu was astoundingly simple. He prepared only a few things, but he did them really well.[2] This approach to quality became the In-N-Out signature. It made the business stand out and quickly built a strong reputation. That was the start of its Noticing Power.

Even today, nearly 70 years later, the food is still prepared in front of you. You can see people cutting the potatoes, slicing the onions, and preparing the lettuce in plain sight. It's simple and honest. And that approach extends to the way the owners conduct their business. They've never floated on the Stock Exchange. They've never franchised the brand. They've barely even touched the original menu (even though they've added a "secret" menu for those in the know). Their family-owned approach reinforces not only the food's authentic local quality but also the company's Staying Power. This forms In-N-Out's brand DNA and is the very soul of its **Iconic Brand Language**™.

This strategy has really worked for it. A few years ago, San Francisco granted the owners permission to open a restaurant in the historic Fisherman's Wharf neighborhood. No other fast-food brand had ever been allowed to do so. This was simply because In-N-Out reflected the family-owned, authentic aesthetic that the area sought to protect.

Interestingly, the owners' approach to **Scaling Power** has run counter to what most people would expect. They've not opted for the explosive international growth that most successful companies go for. Instead, their approach to **Scaling Power** has been slow and steady. This reflects their family-run values as opposed to the profit-is-everything approach of big business. Their unwavering focus on quality means they'll never open a restaurant more than 300 miles from one of their existing food stores. As a result, you can still find stores in only six states.[3]

But their small-town approach doesn't mean they have small-time profits. Revenues are currently more than $500 million a year, and the company has an estimated value of just over a billion dollars.[4]

And it all started with a burger shack smaller than the average garage.

Your Handy Guide to Iconic Advantage®

Now that you understand your company DNA, and you've spent time identifying or developing your signature elements, it's time to articulate and define them in your Iconic Brand Language. Because they're worth nothing to you if all your employees go off and do their own thing regardless.

You might be thinking, "Oh, hold on. We've already got a brand guidelines document that does this." And you'd be partly right. A brand guidelines document does half the job for half the elements. It's focused on controlling everything that makes up the brand—the logo, the colors, the fonts, the tone of voice, the image styles—but we'll be going further to protect the product itself. And we'll be aiming to inspire as well as protect.

This isn't a document that belongs to any one department. We're going to be creating a tool that everyone should use across the business to inform decisions. So you need to make it easy to understand—and maybe think about adding specific sections for different departments to help people in those departments understand how to apply the thinking to their role.

This document, which was inspired by the Visual Brand Language created by one of BMW Group's subsidiaries, Design-works, will be central to building your **Iconic Advantage®**. The better you make your Iconic Brand Language, the more successful you'll be.

Tell the Whole Story

Your product doesn't live in a vacuum. It's likely to exist in an important context that will affect your decisions—especially if you work in a large organization. Your product is probably brought to you by a brand. And maybe even further, it has to fit in with the strategy of a wider holding company.

For example, Dove's men's deodorant sticks sit under the Dove brand and also under the umbrella of Unilever. Diet Coke is a product under the Coca-Cola brand, which is owned by the Coca-Cola Company. All this needs to be understood and considered.

You need to recognize the impact this context has on your product and the decisions you make. To help you understand that, we recommend you start by filling out this template.

Iconic Brand Language™

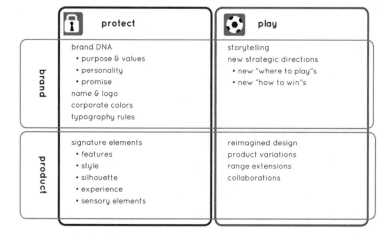

	protect	play
brand	brand DNA • purpose & values • personality • promise name & logo corporate colors typography rules	storytelling new strategic directions • new "where to play"s • new "how to win"s
product	signature elements • features • style • silhouette • experience • sensory elements	reimagined design product variations range extensions collaborations

As you can see from this, you need to start by understanding all the elements of the brand. This is pretty much what traditional brand guidelines do, but—as you'll see in a moment—we want you to take that a step further.

Once you have that foundation, you need to define the elements that are specific to your product. These include attributes like the style, the silhouette, the sound, the texture, the smell, the sensation, and the interaction. These are the elements that form your Noticing Power. Alongside these, you should list the elements that form your Staying Power, which might include your heritage, story, point of view, spokesperson, and purpose.

Once you've listed everything, you need to start defining how each element is used.

Protect and Play

The role of an Iconic Brand Language document isn't simply to restrict people. A frequent complaint about brand guidelines is that they shut down creative opportunities rather than open them up. Your Iconic Advantage depends on creative ideas to thrive, so you want a document that people will find inspiring.

For each element, you need to outline the restrictions *and* the opportunities to play. So instead of just focusing on the barriers, you describe the fertile area contained within the barriers. You excite people with the creative opportunities within a defined area.

For example, the Apple logo is always centered and tastefully proportioned on its products. But there's still flexibility. On some iPhones the logo is an area of polished metal on a matte metal surface. On MacBooks, it's a piece of white plastic that glows when the computer is powered up. On the Apple TV box, it's recessed into the black plastic casing. There's a level of consistent control but also opportunities for creative interpretation.

Understand where the opportunities for expression are for each signature element and make these as clear as the restrictions.

Brand at Play

You're probably familiar with the Chiquita banana logo. (You may even have some of the stickers attached to your monitor at work!) It's recognizable because it's consistent. It is always oval, uses the same condensed font for the name, and usually features the line drawing of the lady with the fruit on her hat.

But there are also slight variations, depending on whether you're looking at a Chiquita "fresh & ready" banana, a plantain, or a Chiquita Jr.

In the early 2000s, the Chiquita brand ran a competition for members of the public to develop new designs for the logo[5]. Of course, this wasn't just an open-ended brief. That would have been a brand disaster. Instead, the company carefully defined what could be played with and what couldn't be touched. That's exactly the kind of understanding every company needs to have about their brands and products.

An Example of Iconic Brand Language™

It will be easier to understand our approach to Iconic Brand Language if we give you an example. So here's a simplified version of what we believe an Iconic Brand Language Matrix would look like for Converse. The company has done particularly well at protecting and playing with its signature elements to keep the product both fresh and consistent.

Iconic Brand Language Matrix™ for Converse

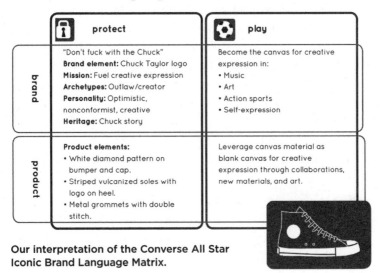

	protect	play
brand	"Don't fuck with the Chuck" **Brand element:** Chuck Taylor logo **Mission:** Fuel creative expression **Archetypes:** Outlaw/creator **Personality:** Optimistic, nonconformist, creative **Heritage:** Chuck story	Become the canvas for creative expression in: • Music • Art • Action sports • Self-expression
product	**Product elements:** • White diamond pattern on bumper and cap. • Striped vulcanized soles with logo on heel. • Metal grommets with double stitch.	Leverage canvas material as blank canvas for creative expression through collaborations, new materials, and art.

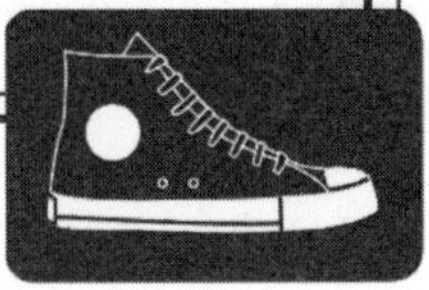

Our interpretation of the Converse All Star Iconic Brand Language Matrix.

What to Protect

We'll start with the brand-level elements. These are probably outlined in a brand guidelines document. But you need to make sure you define them in a way that works for your Iconic Brand Language guide.

Name

If your product exists, it will most likely have a name. If not, this is an opportunity to connect with your audience.

Calling your product the XPR-588P sends the message that you're more interested in the timesheet system of your R&D department than you are in your audience. Having a name that's easy to remember and say is obviously an advantage.

With all of these things, consistency is key. If you're creating variants and product lines, you need to think about defining the rules for your naming conventions as well as protecting the names themselves. And make sure you don't miss the opportunity to use the name to build emotional connections with your consumers and give them any additional information they may need to make a purchase decision.

It's unlikely you're going to want anyone to play with the name. So your job here is to define and protect it. This is one of your product's most fundamental signature elements.

Logo

Every brand guidelines document we've seen focuses on controlling the use of the logo. It dictates colors, size, exclusion zones, and other usage rules. This is important for the media you control, but in this digital age, you need to think bigger.

No matter how you try to protect it, the public will take your logo and use it without your permission in whatever way they like. You have to be OK with that. In fact, more than that, you should embrace it.

The only valuable place a brand exists is in the minds of your customers. And brand guidelines don't apply in there.

In 2011 Coca-Cola went so far in relinquishing control of its logo that it allowed people to replace it with their own names. The names were printed in the Coke style, and the rest of the signature elements remained—but this was revolutionary.[6] Several years later, many

140

people still have a Coke bottle with their name on it.

The only valuable place a brand exists is in the minds of your customers. And brand guidelines don't apply in there.

Another company that's famously flexible with its logo is Google. It has turned the logo into a playground, and you just don't know what you're going to see on the site each day. Sometimes it's the standard logo, sometimes it's an illustration celebrating the birthday of a long-dead artist, and sometimes it's a fully interactive video game.

With both of these examples, there is still consistency. The behavior is consistent. All the other elements you expect are untouched.

So there may be an opportunity to move away from the command-and-control approach and define a behavior that's in sympathy with your product or brand. Just make sure you define it clearly so that it doesn't go off-track.

Colors

Most brands have a well-defined palette of corporate colors. And in most sectors, they tend to be pretty homogenous. Businesses seem to think it's more important to associate themselves with the conservative, trustworthiness of a deep blue than to be noticed.

Very often, corporate color palettes can end up being too restrictive and make companies feel like there is only one side to their personality. If you can, define an area to play here. Under what circumstances would it be OK to use different colors? And what colors are absolute no-nos? Put it all down in writing to protect another important signature.

Typography

Again, this tends to be an overly restrictive section of a brand guidelines document. Your typography is the tone of voice for your communications. If this is too restrictive, it can make you appear one-dimensional.

Just like the use of colors, it's a good idea to define when you can break out of the typography handcuffs. This can open up a world of creative possibilities.

Tone of Voice

Not all brand guidelines have a tone-of-voice section, but it has become more popular over the past couple of decades. It's there to make it seem as if all the words are coming from the same mouth. And it's a vital part of all corporate communications.

Just make sure that there's latitude for playfulness. Out of all the elements, you need to have the ability to adapt here. If you're hoping to connect with individuals on an emotional level, you need to be able to break out of corporate-speak. This is why "authenticity" has become such a popular word in recent years.

But you also need to go beyond words to describe actions. What you do is far more powerful than what you say. Just like body language, your organization's behavior says more than words ever could.

So don't define just the words you can use; define the principles that guide how the brand behaves. That will be far more useful than any set of "if this, then that" mandates.

There may be more things you want to cover at the brand level. These are just the most common and obvious areas. Whatever you define, just make sure you also define the opportunities to play alongside the rules and limitations.

These brand-level factors still play an important role in your Iconic Advantage. Using them with consistency helps to build familiarity and trust. And depending on how you use them, they can also become important signature elements. But, as we've pointed out, you can't stop at the brand level.

So let's look at the important elements of your product and their usage rules. These are some of the things you should probably consider.

Signature Elements

Your signature elements are the very foundation of your Iconic Advantage. So you need to make sure you protect them. Their real power is unlocked when they are solid and consistent over time. That doesn't happen by accident.

If you've done enough work on your Noticing Power, you'll have identified or created desirable points of difference for your product or service. These can be functions, design elements, unique interactions, special moments, processes, and anything else that makes you stand out in a beneficial way.

Always keep in mind the reason why you're doing all of this. It's to provide familiarity. The more familiar people are with your signature elements, the greater your Iconic Advantage.

Highlight the must-have elements—the ones that are the key differentiating points of your product. Now define each of these one by one.

If they are design elements, state exactly how they should be used: the position, the color, the size, the material, the texture, or whatever other attributes you don't want tampered with. Make sure your descriptions are clear and easy to understand for the newest of designers. Don't leave anything as an assumption.

The more familiar people are with your signature elements, the greater your Iconic Advantage.

Of course, your signature elements may not be physical. They could be something less tangible, like a sound, smell, taste, or point of view. But even if you can't touch them, they still need to be protected.

For services, it's equally important to define at which points in the customer journey the signature experiences might occur. You need to identify the signature elements of your interface, user experience, and any other touchpoints. It's just as important to be consistent with these intangible elements as it is with physical ones.

Now define the latitude for interpretation. What attributes of the signature element can be played with? The color? The position? The size? The material? The volume? The smell? Give people some direction on exactly how they can keep the product fresh and interesting. Otherwise, they won't touch it and it will eventually become stale.

After Steve Jobs returned to Apple in the '90s, the company found its design mojo. It pruned its product lines and it developed a design strategy that successfully gave everything an Apple feel. It continues to stay firm and true to these rules 20 years later.

For example, you'll struggle to find a right angle on an Apple product. Everything has rounded corners—from the trackpads on the laptops to the power button on an iPad to the Buy button underneath a Beyoncé album on iTunes. Apple refers to these rounded-edge designs as "lozenges."

It also pays special attention to the most-used interaction points on its devices. It calls these "jewels" and designs them to be extra special—like the Home button on an iPhone, the Digital Crown dial on an Apple Watch, and the trackpad on a MacBook. This extra attention to detail is exactly what's missing from their competitors' products.[7]

These design elements weren't just arbitrary. It wasn't simply a case of saying, "No one else is doing this, so that means we should." Instead, it was something that came from an understanding of the company's DNA and what made it different in the market. Apple focuses on user experience, quality, and design finesse. It defines itself as a trailblazer, breaking new ground and challenging conventions. And these signature elements are a personification of that.

You may never have noticed them before, but now you won't be able to help seeing them in every Apple product.

What to Play With

One of the things that differentiate an **Iconic Brand Language** document from other branding documents is the element of play. **Iconic Advantage** involves confidently walking the line between old and new, consistency and innovation.

Including the flexibility to play is vital. It's what will keep your products fresh, relevant, and exciting. It's what will help you grow your **Iconic Advantage** over time.

Here are some of the benefits it can offer you.

Heritage and Story

Everyone loves a good story. The world's entertainment industries exist because of it. And for many products, their story is what makes them valuable. That's because a story gives meaning.

Think about why people buy souvenirs. No one thinks, "I really must have that overpriced, miniature, mass-produced model of the Leaning Tower of Pisa. It will go beautifully with the interior design theme of my living room." They get it for the story. It says, "I visited this thing. It makes me a slightly more interesting person. I can tell you about it if you want."

You can use the power of story to make your product more iconic.

The story of the Dyson vacuum cleaner is one of a visionary inventor who created better vacuum cleaner technology. He went around to every manufacturer to see if they wanted it. They all said no, so he made the

Iconic Advantage involves confidently walking the line between old and new, consistency and innovation.

product himself.[8] Knowing this story, we imbue the product with those visionary, idealistic, and tenacious characteristics. That makes it more valuable to us than the run-of-the-mill vacuum cleaner designed by an anonymous individual in a factory in Germany. And it allows Dyson to create more of a backstory. This powerful backstory then helps users identify with the product, its founder, and the cutting-edge technology behind it.

Every product has a story of some kind. It may be the story of the founders, the inventors, the designers, the customers, the process, or some historical event it played a part in. Do your research and write the story you want the product to be associated with—preferably one that adds some kind of value to your product.

To do this well, you need to define the core of your story. Don't think of this as a linear narrative; that would become inflexible and shut down creative opportunities. Instead think of the story elements: the protagonist, the challenge, the scenario, the relationships, the moment of truth, the mission, the resolution. Articulate and protect these elements. Then you can tell the story in a myriad of fresh ways. You can zoom in to details. You can pull out to tell the entire tale. You can expand on characters. You can tell it from a different person's point of view. You can set it in a fantasy future. The opportunities are endless, and each tale you tell strengthens and grows the product story.

This additional layer of meaning and relevance helps you build a stronger connection with your audience. And that's central to building Staying Power.

Your **Iconic Brand Language** document isn't useful just at the start of your **Iconic Advantage** process. Its value grows over time. It becomes a guiding light as you develop your product, making it stronger and more relevant year after year.

New Strategic Opportunities

If you refer to our protect-and-play matrix, this is about the top-right quadrant, which is focused on how the brand can play. These are new strategic growth opportunities. They're the "where to play" areas that include opening up new consumer segments, channels, geographies, and product categories.

Your **Iconic Brand Language** document should outline these opportunities clearly. This will inform the designers, marketers, R&D teams, and everyone else where you want to take the brand in the future. It will give them direction on how to marry the protected brand elements with new strategic opportunities.

Reimagining Design

This is where you focus on your product's evolution. The "protect" side of your matrix will help you stay on track and make sure your new product iteration looks as if it's part of the family. But the play side is how you direct your designers' efforts to the right places. You're showing them the most fruitful areas to focus their creative thinking.

Over time, this guidance will help each iteration of the product look like a natural evolution, building on the strong foundation of the past.

Variations, Extensions, and Collaborations

Your **Iconic Brand Language** document also plays an important role in building your Scaling Power. In the last chapter, we looked at the opportunity to tier up and down, extend your signature elements into new products, and create collaborations with other brands. These can be fantastic ways of unlocking new revenue opportunities while building your **Iconic Advantage.**

But you need to do it with sensitivity. It's very easy to damage your **Iconic Advantage** if you do it wrong.

On a wider level, you need to make sure your activities fit with the company's purpose and beliefs. If you betray that for the sake of more income, you'll lose all credibility in the minds of your customers and destroy the connection you've been trying to build.

You also need to make sure your signature elements remain intact and that they don't get obscured. In fact, it's best to use variations, extensions, and collaborations to draw even more attention to your signature elements. That gives you the dual benefits of boosting your **Iconic Advantage** while opening up new revenue streams.

These points are especially important when it comes to brand collaborations. Like any other business deal, they'll often require negotiation and compromise. But it's better not to collaborate than to compromise on your signature elements.

Why an Iconic Brand Language Document Is Worth the Effort

The days of the five- or 10-year business plan are over. The world changes too quickly, and rigorously adhering to a long-term plan will lead to your making inappropriate decisions based on outdated assumptions. You'll miss the new opportunities, because you'll be pushing forward blindly to your imaginary Shangri-la. The companies that follow a rigid 10-year plan are increasingly unlikely to survive to see the end of that period.

These kinds of plans are focused on a defined destination. We think it's just as important to be focused on the journey. Knowing the best decisions to make right now is more important than following instructions written in the past.

A good Iconic Brand Language document gives you the confidence to make decisions in real time. This, in turn, makes you more agile and allows you to take advantage of opportunities while they're still fresh. You'll have a safety net that protects you from damaging the brand in the midst of all the chaos.

It also has significant long-term value. An Iconic Brand Language document keeps you grounded in your iconic heritage so that everything you do will build upon everything you've done before.

It's a guide to making the decisions that will help you build your Iconic Advantage right now and into the future. It's there to prevent your diluting your Noticing Power as you grow. It's there to help you stay relevant as the world changes around you. And it's there to help you scale without losing the magic that makes your product so special.

In an uncertain world, this is your passport to the future—as well as an indispensable guide to the present.

CHAPTER 8

Activating Iconic Advantage®

When Red Bull hit the shelves in 1987, it wasn't just the launch of a new product; it was the launch of an entirely new product category. Energy drinks hadn't really been a "thing" before, and the product stood out on the shelves with its skinny can and bold look. This product attracted a heap of attention.

But just as important as launching a new kind of drink, the company launched a new approach to business that went way beyond a marketing strategy.

It still used traditional media to reach its audience. But it identified a niche market that embodied the spirit of its brand: the extreme sportsperson. Because who needs stimulation and energy more than they? By supporting such an aspirational bunch of daredevils pushing the boundaries of what was possible, the brand was seen every time their antics hit the headlines.

It was phenomenally successful.

The vision and leadership to make this happen came from the very top. Dietrich Mateschitz founded the company with his life savings when everyone thought he was crazy to do so. He understood that branding was absolutely central to the success of the product, so he started working with his old school friend Johannes Kastner to develop the design, tone, and communications.[1] That relationship has remained throughout.

> This rock-solid capability has helped build a level of consistency that few companies achieve. The brand has never wavered from its original purpose and edgy character. It has always been clear on its purpose, understood its audience, and remained focused even while the business developed in unexpected directions. The company's unique point of view became Red Bull's iconic signature element.
>
> Getting the right team in place, empowering them with the right tools and resources, and having them all pulling in the right direction are what has given this company wings. And that's what has made it iconic.

The last chapter was all about creating an Iconic Brand Language™ document. That's absolutely vital. But it won't have any effect unless you bake it into the business and have the right capabilities in place to activate it. That includes your structure, systems, and staff. If you don't, your Iconic Brand Language document will just gather dust on a shelf and nothing will change.

You'll be left holding the architect's plans for a palace while people keep building the modest little cabins they've always built.

It's not just about dreaming it; it's about doing it.

You need to get the right capabilities in place. You need to define your business's growth plans. And you need to extend all of that thinking to your corporate portfolio management.

Fortunately, it's not complicated or particularly onerous. But if you want to get good results, you can't cut corners. The key actions in activating an Iconic Advantage® strategy are collecting the right capabilities; creating a business plan focused on building Noticing, Staying, and Scaling Powers; and finally, at a corporate level, managing your portfolio to maximize iconic value. In this chapter, we'll outline everything you need to put in place and describe the different steps in creating and managing your iconic growth.

Capabilities Required

According to global innovation firm Doblin[2], there are four components that make up a business capability. And they're a great framework for building your Iconic Advantage capability.

These are:

1. Organizational Design
2. Approach
3. Competencies and Resources
4. Metrics and Incentives

The only way to make a dramatic difference in your organization is to address all of these areas. And you need to make sure that they work together to create an effective ecosystem.

Of course, this isn't a small task. But you can't expect to achieve dramatic results by making only modest adjustments. It's best to get the right stakeholders on board and work together to create a sustainable and healthy capability. Let's look at how to do that in more detail.

1. Organizational Design

Get Support from the Top Down

The aim of Iconic Advantage is to transform your business. And you can't do that without having the right people on board.

Clearly, this should start at the most senior level. The most iconic products have the support of everyone from the CEO down. The companies understand the importance of investing in an Iconic Advantage strategy, whether they call it that or not.

Assign Ownership and Accountability

Assign an owner who champions the approach and reports back on the progress. This individual will add clout to the Iconic

Advantage strategy, and everyone working on it is ultimately answerable to this person.

Below this, you will need to staff a team to activate this strategy. This will often include champions from other business functions, because this strategy works best when it permeates every part of the organization. These individuals will be responsible for defining how your Iconic Advantage is brought to life in their disciplines. And that needs to be matched by actions that will help them bring it to life.

It's important that you align your team's goals and incentives with the task of growing and protecting your iconic assets. That demonstrates the company's focus and commitment to developing Iconic Advantage and shows that these individuals have important roles to play in that.

Fund Iconic Advantage

Developing Iconic Advantage is an investment in the future of your business. If you're taking it seriously (which, of course, we're sure you are), you'll assign a budget to make it happen. It takes time, talent, and resources, which don't come free. The more bullish you are with your commitment to becoming iconic, the more returns you can expect.

There are good financial reasons to make a significant investment. As we've pointed out previously, iconic franchises are vastly more profitable than noniconic ones. By investing in making a product iconic, you are giving it a better chance of becoming disproportionately profitable. Continued investment only amplifies this.

Best of all, Iconic Advantage isn't focused just on long-term returns—you can expect short- and medium-term gains too.

But you can get there only if you put your money where your mouth is.

2. Approach

Systemize Your Iconic Brand Language

Ideally, the team you put together will be involved in creating the Iconic Brand Language we outlined in the last chapter. If they feel an ownership in creating it, there's more chance they'll be motivated to act on it.

This team will be responsible for infusing the Iconic Brand Language at the right moments in every process, from product development to marketing, distribution to finance.

For example, in the design process, it would have a place at the beginning to help ground the design brief, would be available for inspiration during the design process, and would be used at the end to help judge the design concepts.

You would certainly use it to help you judge mood boards, at every stage of product design, during the UX and UI development of your service offering, and at various stages of developing and implementing your marketing campaigns.

You may find that it can also play an important role in candidate selection when you're recruiting, in deciding the right businesses to buy and divest, and in selecting the right retailers to stock your products.

How you use your Iconic Brand Language document depends on your business. But don't just leave it to chance, or it won't be used at all. You need to define where, when, and how you use your Iconic Brand Language to make sure it has the maximum effect across every function.

Remember that your Iconic Brand Language shouldn't be a static document. You should also work with your team to regularly assess and refresh the thinking—because if the world doesn't stand still, you can't expect some one-time thinking to remain relevant indefinitely. These sessions are also good opportunities to discuss what's working, what isn't, and what can be improved.

Keep Your Iconic Brand Language Alive

Most employees are skeptical about new initiatives, and the likelihood is that they'll see the introduction of an Iconic Advantage approach as just another inconvenience or fad. It will probably take a bit of effort to get people behind it.

Don't expect printing out a copy of the document and leaving it on people's desks—or emailing a link to its location on the intranet—to have any impact. You need to explain to the relevant people why this is important and why you expect them to get behind this. Make it relevant to them and show them how it's about opportunities rather than just extra work or disrupting their workflow.

Be a fervent believer and evangelize the message of the iconic promised land to the people you need to get on board. But also keep in mind that changing actions is more important than changing beliefs. As their actions bear fruit, their minds will follow.

3. Competencies and Resources

Gather the Right Talent

You need to make sure you have the right mix of skills in the people on your team. If you want to build strong Noticing and Staying Powers, your team members need to be comfortable marrying the old and the new. They need to understand the importance of protecting signature elements as well as playing with them. They need to understand the past and be focused on the future.

So it's a good idea to seek out the following skills:

▶ **Organizational**
Implementing the approaches described in this book takes work. You need to have people who know how to plan the process and make sure everything gets done.

155

- **Analytical**

  If you don't put KPIs (key performance indicators) in place and keep an eye on them, you'll have no idea if your Iconic Advantage strategies are working for you.

- **Creative**

  An Iconic Advantage approach is hungry for ideas. You need to have people who can focus their thinking in the right directions to create signature elements and keep the product fresh over time.

- **Advocacy**

  You want to get the entire company behind an Iconic Advantage strategy. That involves sharing the vision and opportunities with even the most skeptical employees.

- **Historical**

  You will need people to cherish the past and the learning gained from experience and historical perspective.

- **Directional**

  Someone needs to hold it all together and be ultimately responsible for the success of the strategy. That person may be you.

Manage the Product Life Cycle

If you want your product to remain relevant to your audience, you need to manage its life cycle. The decisions you make in this process are central to its ongoing success.

First of all, you need to create a multiyear line plan. We recommend looking about three years ahead so that you're covering a few iterations of the product. This involves having a clear vision about where the product is going, aiming to make it stronger and more relevant with each design refresh. You need to find the right balance of newness and oldness, freshness and familiarity. That's something that varies from industry to industry and brand to brand. Your aim is to ensure continuity and stimulate repurchasing from your audience.

There may come a time that your product falls out of favor and there's little you can do about it. That could be because it's become overcommoditized or because it is too strongly associated with a moment in time. If that's the case, the wisest decision could be to remove the product from the market and put it into your brand vault. This involves archiving everything you can about the product, from its materials and manufacturing process to its packaging and marketing materials—and, of course, its Iconic Brand Language. There may come a time when you can bring it back to market as a retro revival. It's easier to do that if you have all the materials you need archived in one place.

Find the Right Partners

This isn't just about the internal team. There are some parts of the puzzle that require outside involvement.

Having a strong network of collaborators makes you stronger. The right partners will add to your Staying Power and help broaden your market. Just make sure you have similar values and points of view, or you'll weaken your Iconic Advantage.

If you're choosing a partner to collaborate with on a product or a media partnership, or as an exclusive distributor, this is an opportunity to piggyback on that partner's own iconic strengths. If the company has a strong connection with an audience, it may be able to open up that market for you. Borrowing someone else's Staying Power can be a powerful way of boosting your own.

You may find that some of your partners might actually be your users. Products that connect with an audience will naturally form communities of like-minded, passionate individuals. If that's the case with your existing product, it's probably a good idea to involve some of the most influential members of these communities in some way—because it's easy to alienate your loyal customers if they feel you've made changes in your own interest rather than theirs. Don't overlook the power of your community. They could be the key to your success.

Nike has a rather unusual approach to brand historians. It understands that its brand lives in the minds of its customers. And it knows that, as well as being knowledgeable about the products, sneakerheads are likely to be far more passionate than anyone employed by the company about being the guardians of the brand's history.

So Nike tapped into its community of enthusiastic collectors to form a team of part-time brand historians.

It designates these individuals as a type of brand historian. This may even involve having them turn their garages into miniature Nike museums filled with their prized collections.

These individuals are delighted to give their time to talk or blog about what they love. And more important than the recognition is the chance for them to play a part in the future of a company they worship.

Not every brand can generate this kind of passionate following, but for those that do, this approach can be a powerful way of engaging with their communities.

---

4. Metrics and Incentives

Measure the Important Stuff

There's a popular Peter Drucker quote that applies here: "If you can't measure it, you can't improve it."[3]

If you want to constantly build and improve your Iconic Advantage, you need to know how your Noticing, Staying, and Scaling Powers are doing. Here are some of the measurements you should put in place to track your progress.

Noticing Power You should be measuring this quantitatively by adding a question in your annual brand health tracker. Your aim is to find out how distinctive your product or service is in comparison to the competition.

This should be accompanied by qualitative research to get a sense of how powerful your signature elements are. You can do this with a simple customer panel. See how well these individuals identify your signature elements and find out how much value they place on them.

When you understand your signature elements, you can conduct a side-by-side quantitative discrete choice measurement to see how your product stacks up against the completion.

Staying Power Again, you should add another metric to your annual brand health tracker. This time, it's to measure relevance. You can break this down further to measure familiarity, meaning, delight, and excitement—the four key qualities of Staying Power.

Scaling Power The final metric to add to your annual brand health tracker is aided and unaided awareness. You can extend this more specifically to your key signature elements to find out how well people understand and remember them.

Create A Business Plan and Strategy

Once you've got your team together and everyone's aligned on your iconic ambitions, your first job is to develop an Iconic Advantage business strategy together. There are three main areas you need to work on collectively.

1. The Goals and Aspirations of the Franchise

Set a clear goal to reach iconic status within a defined customer group. Your objective here is to become the standard-bearer for an iconic benefit within your category.

2. Where to Play

Next, you need to define the universe you want to become iconic in. Define the consumer segments, product/service segments, channels, and geographies you wish to become iconic in. We recommend going after a smaller segment first and then building from there.

3. How to Win

Once you have a well-defined universe, actively create a program to win with Iconic Advantage. Outline how you will build greater Noticing and Staying Powers versus your competition.

Develop a Multiyear Growth Program

Now it's time to turn these strategies into a plan. Start by developing a multiyear plan that includes action points, responsibilities, and financial targets. Iconic Advantage can make financial impacts in the short term as well as the long—so it's a good idea to include it in your profit-and-loss forecasts.

We recommend that you plot your capabilities and activities on a timeline and assign a budget to them. Do this over a three- to five-year period, clarifying the key points in your product development cycle, innovation plan, and marketing activities.

Your objective here is to become the standard-bearer for the iconic benefit within your category.

Develop a Design Plan for Greater Noticing Power

The first step in your design plan is to create and maintain Noticing Power. Take steps to make sure you have strong signature elements and that you are constantly making them stand out and be noticed. These are the crown jewels that your designers need to constantly celebrate.

Next, it's important to keep your product fresh if you want to stay ahead of the competition. That requires a multiyear design plan. How often you refresh your product depends on the industry you're operating in. But as soon as you stand still, you leave yourself open to attack. And, in today's fast-moving market, you can become irrelevant overnight.

► Work out how often you need to refresh your product. If in doubt, err on the regular side. Schedule these refreshes into the workflow using your Iconic Brand Language as your guide and inspiration.

► Consider partnerships too. Collaborations and special editions can be a great way to keep your offering fresh and relevant and to continually expand your audience.

Develop an Innovation Pipeline to Maintain Staying Power

You also need to think about developing the product beyond its design aesthetics. Your signature elements should be a work in progress. You should regularly work on them to make them more powerful over time.

Nike has done this brilliantly with the soles of its Nike Air sneakers. The air pocket has been developed from a small patch in the heel of the original Air Max shoe to the entire sole of the current design. This has made the benefit more powerful and increased the visibility of the signature element.

The best companies go beyond innovating on their signature elements; they innovate on the benefits the signature elements deliver. These products rise above the market to become standard-bearers for the category benefit.

In Chapter 4 we shared the story of Amazon's relentless pursuit of owning the benefit of "no patience required" with its 1-Click approach. This total focus on owning a benefit, even at the expense of cannibalizing its existing businesses, is what will ensure that Amazon stays timelessly relevant.

Evolving the signature elements in this way widens the gap between your product and the alternatives. It makes it harder to copy you. And it keeps your customers interested (and continuing to swipe their credit cards).

As well as innovating on your signature elements, you should also consider using your signature elements to develop new, innovative products. As we described in Chapter 5, this can include tiering up, tiering down, and collaborating to open up new markets and opportunities for your existing product. But there may also be an opportunity to develop adjacent product lines. Just make sure that everything builds on your Noticing and Staying Powers to strengthen the connection with your audience.

Develop a Marketing Program to Enhance Staying and Scaling Powers

As we've mentioned, Iconic Advantage puts the power of marketing into the product itself. So the purpose of advertising and promotions is simply to draw attention to the signature elements and avoid borrowed interest and gimmicks. You'll want to create a multiyear program that will help you do this consistently over time without getting stale.

Your story and your purpose are part of your signature. These should already be outlined in your Iconic Brand Language document. You want them to be consistently understood inside and outside of the company. But that doesn't mean you just repeat the same thing again and again. That would just make you an unimaginative bore (which you're clearly not).

Think of your brand story as being like Tintin, the main character of a long-running comic series. He remains absolutely consistent throughout the many adventures he has. Your brand story should be like that. Come at it from different angles to communicate a consistent message. And make sure your signature elements sing out from every medium.

This takes strong leadership. If you're working with agencies, you need to brief them well, focusing on the opportunities to play rather than the restrictions you're imposing. And choose your media wisely—it can say as much about you as the message you put in it.

Don't fall for the typical shouty sales message approach. Good signature elements differentiate your product in a way that's relevant to your audience. Your marketing should be about getting your product in front of people's eyeballs in an appropriate way.

Spend time developing a plan. Work out where you want to be and by when. And measure progress to make sure you stay on track.

If you've ever read a Jack Daniel's ad, you've probably got a pretty good idea of what Lynchburg, Tennessee, is like. It's an old-time, one-horse town that's kept alive by the Jack Daniel's whiskey distillery. Things are done in the same way that cantankerous old Jack did them back in the 1800s. And many of the people who work in the distillery look as if they came from that era.

Of course, it can't possibly be an accurate account of the situation. The distillery has recently undergone a dramatic expansion to be able to fulfill an increasing demand for Jack Daniel's whiskey. And the company has extended its product line to some interesting variations that have opened up new markets and sales opportunities.

But the consistent story told through the black-and-white ads is one of traditional values and small-town spirit—with the iconic product acting as a full stop at the bottom right-hand corner of every piece. That's a conscious and wise decision. Every piece of communication simply adds to the mythology and reinforces the brand character.

Jack Daniel's' advertising agencies have proved that it's possible to tell a consistent brand story in fresh and interesting ways, year after year. Like the contents of the iconic bottles, this is a formula that would be foolish to change.

JACK DANIEL'S SEVEREST CRITIC is our whiskey taster who makes sure our whiskey is gentled to the proper sippin' smoothness.

This gentleman's word is law at Jack Daniel's small distillery. He tastes our whiskey just as it comes from the room-high Charcoal Mellowing vats. And if he should say "no," the whole batch would be rejected, and the charcoal replaced for the next run. You see, we're not taking any chances on changing the quality of Jack Daniel's. One sip, we believe, will tell you why.

THE TENNESSEE ◊ SIPPIN' ◊ WHISKEY

TENNESSEE WHISKEY • 90 PROOF BY CHOICE
DISTILLED AND BOTTLED BY JACK DANIEL DISTILLERY • LYNCHBURG (Pop. 361), TENN.

RIGHT HERE, IN THIS VAT of hard maple charcoal, is where Jack Daniel's becomes a smooth, sippin' Tennessee Whiskey.

When we first make Jack Daniel's it's much like any good whiskey. But then, in our mellowing house, we give it an extra blessing. Here, every drop is seeped through twelve feet of charcoal before aging. And this slow trip puts it in a class all its own. Charcoal mellowing is why no other whiskey achieves such rare, sippin' smoothness. And why our labels will always read: Jack Daniel's *Tennessee Whiskey.*

CHARCOAL MELLOWED ◊ DROP ◊ BY DROP

TENNESSEE WHISKEY • 90 PROOF
DISTILLED AND BOTTLED BY JACK DANIEL DISTILLERY • LYNCHBURG (POP. 361), TENNESSEE

AT JACK DANIEL'S DISTILLERY, these gentlemen make whiskey as their fathers before them did.

There are dozens of men who work in our Hollow whose fathers have worked here too. That's good, because it means the elder hands can pass their knowledge and skills to newer generations coming along. That's the way it's been since Mr. Jack Daniel taught his nephew, Lem Motlow, how to make whiskey. A sip, we believe, and you'll be glad we still set store in family tradition.

SMOOTH SIPPIN' TENNESSEE WHISKEY

Jack Daniel's adverts from 1961, 1973, and 1987.[4]

Develop an Iconic Portfolio Management Practice

One of the big mistakes in business thinking in recent years is the focus on milking cash cows and overinvesting in rising stars. It's not that you shouldn't invest in new products. In fact, you absolutely should. But starving and neglecting your cash cows lead to missed opportunities. Iconic franchises yield disproportionately higher profits, so we think it's wiser to invest in developing them than milking them dry.

Instead of looking only at growth and market share as the key qualities when making portfolio decisions, we recommend using the following Iconic Advantage Portfolio Matrix.

Iconic Advantage Portfolio Matrix™

	low iconic potential	high iconic potential
high revenue growth	Develop stronger Noticing and Staying Powers	Double down investment against Scaling power while maintaining a strong investment in Noticing and Staying Powers
low revenue growth	Divest	Invest more heavily in Scaling Power within a niche

The two axes that we're looking at are revenue growth and iconic potential. The aim is to move to the top-right quadrant, where you've maximized your iconic potential and it's paying off with profits. Let's examine each quadrant in more detail.

High Potential and High Growth

If you've got a product that is already successful and shows a good chance of even more success, you're very lucky indeed! Your job is simply to amplify the Iconic Advantage that the product already has.

Most of this investment will go into developing its Scaling Power. But as part of its life-cycle management, you also need to keep developing its Noticing and Staying Powers. This is the quadrant everyone longs to be in.

High Potential and Low Growth

If you have a product that has potentially strong Noticing and Staying Powers but have not been able to scale it, this is your quadrant. Your job here is first to determine the most important niche to become iconic within. Then focus your efforts on doubling down on Scaling Power within that narrow universe. Only after you win over that niche's audience should you consider expanding your scaling universe.

High Growth and Low Potential

This is for products that have been successful but lack strong Iconic Advantage. Your job here is to noticeably differentiate your product from the competition and help your audience connect with it on an emotional level. We recommend that you work on creating signature elements that will help to create valuable differentiation. Once you have these in place, you should invest in Staying Power. If you do that successfully, you'll appeal to a greater portion of the market and increase your chances of gaining long-term repeat business.

Low Growth and Low Potential

This is a tough one, especially if you or the business has an emotional attachment to the product. But if it's not doing your business any good, you need to get rid of it. You can either divest it or put it into your archive. Either way, you sometimes have to admit that it's not worth throwing good money after bad.

Use this framework to assess the products across your business. You may want to do this in a team to get a variety of viewpoints. Then focus your efforts in the right place to create the biggest impact.

Buckle Up and Let's Get Going

Iconic Advantage starts with a destination—the place where your product will be the first option your audience thinks of in your category.

Then create a strategic road map laying out how you'll get from where you are now to this place of iconic leadership.

Finally, you need a working vehicle with a powerful engine, lots of fuel, a spare tire, and a competent driver to follow the road map. That's what your strategic capabilities are. You need to have the staff, processes, resources, and corporate portfolio management practice in place to bring your Iconic Advantage to life.

Without the right capabilities, your iconic ambitions will remain just a pipe dream.

CHAPTER 9

Final Thoughts

As you can see, this isn't a particularly complicated strategy. It starts with an ambition to have the most iconic product in your category. And then there are three steps to follow:

1. Build greater **Noticing Power** by developing signature elements.

2. Enhance **Staying Power** by creating timeless relevance—both functionally and emotionally.

3. Create **Scaling Power** through your product extensions, marketing, and distribution.

These steps increase the probability of becoming iconic and staying iconic over time.

They apply equally to existing iconic franchises, existing franchises that have the potential to become iconic, and new businesses that are still in development. And they're as relevant to services as they are to products.

Iconic Advantage Framework™

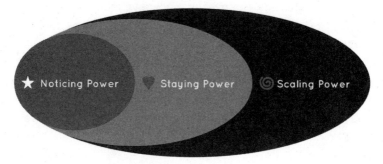

Giving your New Business the Right Foundation

If you're in the planning stage of a new business, starting with an Iconic Advantage® strategy will give you a great head start. It's easier to begin with strong Noticing and Staying Powers than it is to add them afterward. It will help your offering stand out against the competition. It will ensure that you're relevant to your target market. And baking the marketing into the product will help your advertising work harder, even on a tight budget.

This is one of the best things you can do to give your new business the best chance of success.

And We Do Mean "Advantage"

We call this strategy Iconic Advantage for a very good reason. It's about giving your product the edge over the competition. Noticing Power helps you grab a higher share of attention in your market. And Staying Power helps you be more relevant to your audience than your competition is.

This is especially important in industries with a multitude of generic offerings. But even in innovative industries, like tech start-ups, it can give you that vital boost in a two- or three-horse race. The reason is because people tend not to know what the number-two picture-sharing social media site is. Or the second-most-popular site for renting a room by the night. When you're in an industry where one option dominates the market, you need to be more distinctive and more relevant than the rest.

Stinky cheese beats a fancy mousetrap every time!

The Discipline of Iconic Advantage

Although creating Iconic Advantage isn't hard to do, it takes discipline. But nothing in business gives you a dramatic outcome without commitment and applied effort. The entire strategy can be summed up in this diagram:

Iconic Advantage Strategy

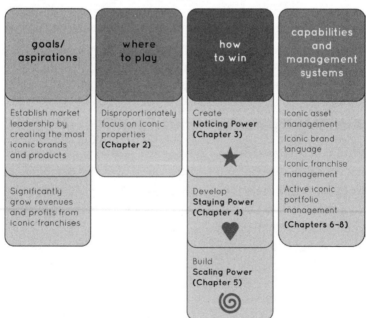

Probably the hardest part of Iconic Advantage is remaining disciplined and staying the course. It's easy to be wowed by shiny objects that can distract you and lead you off-course.

Depending on your organization, you may also have a challenge in getting people to want to work on your heritage product. Your colleagues may see it as a fuddy-duddy part of the business that just doesn't have the same excitement as the launch of an innovative new product. You will need to put effort into selling the bigger vision of why this is important and provide the proper internal recognition and rewards for doing it—because only the best are assigned the family jewels, and it's a rite of passage for anyone looking to lead the larger organization.

Are You Brave Enough to Be a Cannibal?

You can't dip your toe in the water with Iconic Advantage. If you want to make big waves, you need to be all-in. We've found that the organizations that succeed with Iconic Advantage are so religious about owning the iconic category benefit that they're even willing to create new businesses that threaten to cannibalize their existing ones.

For example, the Amazon Kindle has undoubtedly eaten into Amazon's physical book sales. And Apple's iPad is coaxing people away from buying MacBooks. This kind of move is politically difficult to do. The infrastructure and staff of one division will be negatively impacted by the success of a different division.

Naturally, this kind of situation activates the corporate antibodies that will try to kill or inhibit what is perceived as a threat. Fear will become a problem if you let it grow unchecked. It takes strong and unselfish leadership to manage this kind of move. But in the long run, it's the best approach.

Why Pursue an Iconic Advantage Strategy?

Sure, we've already discussed this and given you a myriad of reasons. But there's no harm in exploring this again as we hurtle toward the back cover of the book.

First of all are the financial reasons. This strategy doesn't require a huge expense. You can probably undertake the whole process with your existing staff and infrastructure.

It's also less risky than pursuing only a disruptive innovation strategy. When you innovate on the old, you've already got a product (or service), production, distribution, and—most important—an audience. That's the foundation you're building on. An Iconic Advantage strategy is focused on getting the most value from these existing assets.

Because you are leveraging existing assets, it can help you generate substantially more profit. And it does that by helping you reduce manufacturing costs as you scale, and allowing you to charge more for a product that possesses a historical emotional value for your customers.

Second, it's also a simpler strategy that's easier to implement than building new capabilities. Although it doesn't require many new technical skills to implement (since you are leveraging many of your existing resources), it does require strong leadership and commitment across your organization. The temptation to be distracted by shiny new objects is especially strong for those with the most resources and influence.

Last, if you want to make a mark on your organization and be hailed as a hero, following an Iconic Advantage strategy is a great way to achieve that. You'll make a radical impact on the short- and long-term success of the business. You'll elevate your company's offering above the competition's in a more sustainable way. And that's an outcome everyone in the business wants.

These are all compelling reasons to pursue an Iconic Advantage strategy: you'll make more money, it's easier to do, and it will help your career. However, the real reason to pursue this strategy goes much deeper.

It's About People

There was an outcry when it was announced that Twinkies were being discontinued—not necessarily because people were regular consumers of the product, but because the product had a place in their heart. It was part of people's rose-tinted past, and the idea of not being able to relive that part of their personal history (or have others share in this experience in the future) was unthinkable. Iconic brands have the power to make people feel something, and that's what makes them special.

This book has hopefully helped you understand the importance of iconic brands, because they represent a part of who we are when we use them. If you have read this far, it likely means there's a brand you care enough about to want to elevate it to an iconic level.

That's good for the brand. But it's also good for all of us.

Each of us, whether we admit it to ourselves or not, has a love affair with a number of brands. As with all good romances, we don't want it to end. We want to keep that love in our lives. We want to be all-in.

As the caretaker for your brand, it's your job to keep that love alive.

We hope you keep developing Iconic Advantage for your own products. And we hope, in the process, that you spread a little more happiness and meaning throughout the world.

Acknowledgments

To our colleagues, friends, and family, who have been so patient with us while we ventured on this journey of self-discovery and doubt, new friendships and collaborations, and endless cycles of satiated curiosity followed by the inevitable humility of ignorance: We thank you for your love and support.

To our brave content advisors who read our roughest proposals and manuscripts, **Judy Chen, Xia Feng, Ann Kositchotitiana, Karina Krulig, Anand Muralidaran, Bhavesh Naik, Kim Wang, Simon White,** and **Fiona Zwieb:** You helped us be clearer, more concise, and hopefully more compelling.

A special shout-out to the many thought leaders—**Arne Arens, Lisa Bodell, Jonah Berger, Richard Butterfield, Andrea Cannelloni, Zoe Chance, Alexa Clay, Iain Douglas, Adrienne Eberhardt, Heather Fraser, Jeremy Gutsche, Adam Grant, Andrew Hurteau, Alice Inoue, Rafferty Jackson, Piotr Juszkiewicz, Ardy Khazaei, Parker Lee, Edward Mady, Peggy McAllister, Mitch Murphy, Brian Quinn, Jeneanne Rae, Tom Rath, Dan Roam, Mark Sato, Timo Schmidt Eisenhart, Paul Stead, Chandra Surbhat, Joy Thomas, Dorothea Volpe, Vince Voron, Phil Waggoner,** and **David Wolfe,** and **Garsen Yap**—who pushed us to be better and broader yet more tangible with our ideas.

This book would never have taken shape without the expertise and skills of these amazing collaborators, including the publishing team of **Dabney Rice, Shannon Marven, Liz Wiseman, Nena Oshman, Austin Miller, Anthony Ziccardi, Michael Wilson, Billie Brownell,** and **Sarah Heneghan**; the design and creative inspiration direction of **David Palmer,** and **Paul Stead;** the platform-building talents of **Romina Kunstadter, Kelly Schram, Rachel McDonald, Tara Berthier, Tori Marra,** and **Caitie Bradley;** and the wisdom and generosity of **Richard Nash**.

We are especially grateful to the folks at BMW Group and BMW Designworks, including **Tim Mueller, Birgit Pucklitzsch, Sonja Schiefer, Dee Supavanichy, Peter Falt, Adrian Von Hooydonk,** and **Frank Stephenson,** who helped pave the way for building truly iconic properties that inspire all of us.

For showing us the light when we were lost in the dark, **Philip Hamilton, Christopher Williams,** and **Iris Yen,** the amazing leaders from the Nike family (both past and present).

During this journey, we have been blessed by the continual support and smarts of the Continuum Innovation team, including **Chris Michaud, Gretchen Rice, Liwen Jin,** and **Brian Wen.**

Most important, we would like to thank our mentors who created pivotal moments in our journey by kicking us in the butt when we got stuck, extending a hand when we fell, and bending a gentle ear when we were full of ourselves. Thanks to **Roger Martin,** who first challenged us to write this book because he saw the potential before anyone else did. To **Claudia Kotchka,** for teaching us so much about innovation and people. To **Charles Zook,** for doing the near impossible: keeping us honest with ourselves. To **David Chao,** for always championing the underdogs. To **Stephen Dull,** for letting us be a small part of your amazing legacy. To **Martino Scabbia Guerrini,** for having the courage to marry art with science. To **Bob Shearer,** for always asking why. To **Eric Wiseman,** for having a contagious vision. To **Ray Wang,** for trailblazing without torching the path for others. To **Wen Hsieh,** for having faith in an old mentor. And

Acknowledgments

to **Chip Heath,** who was there from the very start before there was even fertile soil—you helped plant seeds and stuck around even during the droughts.

Last, we want to thank our family members who had faith in us when we left our jobs to pursue new dreams.

Dave wants to thank **Valerie, Iona,** and **Simone.** When the words weren't playing nicely, they proved there was more to life than rattling a keyboard.

And from Soon, thank you to my parents, **Albert** and **Jean,** and sister, **Alice,** for always being rocks of support. And finally, to **Christine** and **Brenden,** for being the tickle in my laugh and twinkle in my smile every day.

Endnotes

Chapter 1

1. Amelia Hill. "The Rise and Fall of American Apparel," *The Guardian*. Retrieved June 1, 2017, from www.theguardian.com/business/2010 /aug/25/rise-fall-american-apparel.

2. Shan Li. "American Apparel Rapidly Grew Its Retail Footprint. Did That Strategy Contribute to Its Collapse?" *Los Angeles Times*. Retrieved June 1, 2017, from www.latimes.com/business/la-fi -american-apparel-stores-20170118-story.html.

3. Charles Riley. "American Apparel Fires Controversial CEO," *CNN*. Retrieved June 1, 2017, from money.cnn.com/2014/06/18/news /american-apparel-ceo-dov-charney/index.html.

4. Mark McSherry. "American Apparel CEO Dov Charney Fired: The Fall of a Merchant of Sleaze," *The Guardian*. Retrieved June 1, 2017, from www.theguardian.com/business/2014/dec/17/american -apparel-ceo-dov-charney-fired-fall-icon-sleaze.

5. Elizabeth Segren. "The Fall of the Hipster Brand: Inside the Decline of American Apparel and Urban Outfitters," *Racked*. Retrieved June 1, 2017, from www.racked.com/2015/3/3/8134987/american- apparel-urban-outfitters-hipster-brands.

6. Phil Hazlewood. "Millennial Mini Straight From Production Line to Museum," *The Guardian*. Retrieved June 1, 2017, from www.theguardian.com/business/2000/oct/05/manufacturing.

7. Silke Konsorski-Lang and Michael Hampe, eds. *The Design of Material, Organism, and Minds: Different Understandings of Design* (New York: Springer, 2010).

8. "MINI: Design DNA," *Car Body Design*. Retrieved June 1, 2017, from www.carbodydesign.com/2012/03/mini-design-dna/BMW.

9. MINI website, shop.mini.com.

10. A. G. Lafley and Roger L. Martin. *Playing to Win: How Strategy Really Works* (Harvard Business Review Press, 2013).
11. Carmine Gallo. "Steve Jobs: Get Rid of the Crappy Stuff," *Forbes*. Retrieved June 1, 2017, from www.forbes.com/sites/carminegallo/2011/05/16/steve-jobs-get-rid-of-the-crappy-stuff/#75bf45b07145.

Chapter 2

1. Discussion between Liwen Jin and Brian Wen of Continuum, June 15, 2017.
2. Alastair Sooke. "Milton Glaser: His Heart Was in the Right Place," *The Telegraph*. Retrieved June 1, 2017, from www.telegraph.co.uk/culture/art/art-features/8303867/Milton-Glaser-his-heart-was-in-the-right-place.html.
3. Dave Walker. "Building Brand Equity Through Advertising," *Ipsos-ASI*. Retrieved June 1, 2017, from www.ipsos-asi.com/pdf/rc5.pdf.
4. Randall Beard. "Make the Most of your Brand's 20-Second Window," *Nielsen*. Retrieved June 1, 2017, from www.nielsen.com/us/en/insights/news/2015/make-the-most-of-your-brands-20-second-windown.html.
5. R. Adolphs R and M. Spezio. "Role of the Amygdala in Processing Visual Social Stimuli," *PubMed.gov*. Retrieved June 1, 2017, from www.ncbi.nlm.nih.gov/pubmed/17015091.
6. Saul McLead. "Maslow's Hierarchy of Needs," *Simply Psychology*. Retrieved June 1, 2017, from www.simplypsychology.org/maslow.html.
7. Mary Fairchild. "Christianity Symbols Illustrated Glossary," *ThoughtCo*. Retrieved June 1, 2017, from www.thoughtco.com/christianity-symbols-illustrated-glossary-4051292.
8. Nigel Hollis. "What Makes an Iconic Brand?" *WPP*. Retrieved June 1, 2017, from www.wpp.com/wpp/marketing/branding/whatmakesaniconicbrand/.
9. Karen Freeman, Patrick Spenner, and Anna Bird. "If Customers Ask for More Choice, Don't Listen," *HBR*. Retrieved June 1, 2017, from hbr.org/2012/05/customers-arent-as-savvy-as-yo.
10. Iconic Advantage benchmarking study conducted June 2014 to July 2016.
11. Michael Porter. "Porter's Generic Competitive Strategies (Ways of Competing)," *University of Cambridge*. Retrieved June 1, 2017, from www.ifm.eng.cam.ac.uk/research/dstools/porters-generic-competitive-strategies/.

Chapter 3

1. Todd Krevanchi. "The Father of Nike Air: Marion Franklin Rudy," *Sneaker History*. Retrieved June 1, 2017, from sneakerhistory.com /2015/01/the-father-of-nike-air-marion-franklin-rudy/.

2. Greg Vernick. "Nike Air Max—A Historical Perspective," *Freshness Mag*. Retrieved June 1, 2017, from www.freshnessmag.com/2015/03 /26/nike-air-max-a-historical-perspective/.

3. Horatiu Boeriu. "How the Perfect Car Door Sound Is Made for BMW," *BMWBlog*. Retrieved June 1, 2017, from www.bmwblog.com/2014 /12/22/perfect-car-door-sound-made-bmw/.

4. Anne Trafton. "In the Blink of an Eye," *MIT News*. Retrieved June 1, 2017, from news.mit.edu/2014/in-the-blink-of-an-eye-0116.

5. "Visual Attention Software (VAS)," *3M*. Retrieved June 1, 2017, from solutions.3m.com/wps/portal/3M/en_US/Graphics/3Mgraphics /ProductsGPIM/Products/~/3M-Visual-Attention-Software-VAS?N =5786+8697840+3293010368+3294529207&rt=rud.

6. Frederick L. Coolidge. "Why People See Faces When There Are None: Pareidolia," *Psychology Today*. Retrieved June 1, 2017, from www.psychologytoday.com/blog/how-think-neandertal/201608 /why-people-see-faces-when-there-are-none-pareidolia.

7. skalka7. "Screaming Wall," *Flickr*. Retrieved June 1, 2017, from flic. kr/p/9kSrjT & Gary Davidson. "Faces in strange places 001," *Flickr*. Retrieved June 1, 2017, from flic.kr/p/5LvqR5.

8. Klaus Heine. "The Concept of Luxury Brands" (2nd ed., 2012). Retrieved June 1, 2017, from www.conceptofluxurybrands.com /content/20121107_Heine_The-Concept-of-Luxury-Brands.pdf.

9. Dyson website, "About Dyson" section. Retrieved June 1, 2017, from www.dyson.co.uk/community/aboutdyson.aspx.

10. Kelly McLaughlin for Dailymail.com and Associated Press, and Zoe Szathmary for MailOnline. "How the Simple but Brilliant Design of the Kikkoman Soy Sauce Bottle Helped Bring the Brand Into Dining Rooms Across the World...and Even Into the MoMA Collection," *Daily Mail.com*. Retrieved June 1, 2017, from www.dailymail.co.uk /news/article-2947710/Kikkoman-bottle-symbolizes-soy-sauce -world-man-created-bullet-train-dies-aged-85.html.

11. Sir Richard Branson. "Virgin Atlantic: 30 Years of Fun, Flying and Competition," *The Telegraph*. Retrieved June 1, 2017, from www.telegraph.co.uk/finance/comment/10917094/Virgin-Atlantic -30-years-of-of-fun-flying-and-competition.html.

12. Ashley Rodriguez. "How Allstate's Mayhem Disrupted the Chatter Around Insurance," *Ad Age*. Retrieved June 1, 2017, from adage.com

/article/cmo-strategy/mayhem-helped-allstate-disrupt
-conversation-insurance/298779/.

13. LEGO website, "LEGO History Timeline" section. Retrieved June 1, 2017, from www.lego.com/en-gb/aboutus/lego-group/the_lego _history.

Chapter 4

1. Google entry on Wikipedia. Retrieved June 1, 2017, from en.wikipedia .org/wiki/Google.

2. Google website, "Our Products" section. Retrieved June 1, 2017, from www.google.com/intl/en/about/products/.

3. "How Much Search Traffic Actually Comes From Googling?" *eMarketer*. Retrieved June 1, 2017, from www.emarketer.com/Article /How-Much-Search-Traffic-Actually-Comes-Googling/1011814.

4. Jacob Gube. "Popular Search Engines in the 90's: Then and Now," *WebpageFX*. Retrieved September 1, 2017, from www.webpagefx. com/blog/web-design/popular-search-engines-in-the-90s-then -and-now/.

5. Adam Boome. "How Food Superbrands Manage to Become Your Family," *BBC News*. Retrieved June 1, 2017, from www.bbc.co.uk /news/business-13598581.

6. Chris Anderson. *The Long Tail: Why the Future of Business Is Selling Less of More*, rev. ed. (Hachette Books, 2008).

7. Jim Camp. "Decisions Are Emotional, Not Logical: The Neuroscience Behind Decision Making," *Big Think*. Retrieved June 1, 2017, from bigthink.com/experts-corner/decisions-are-emotional-not -logical-the-neuroscience-behind-decision-making.

8. Josh Halliday. "Gap Scraps Logo Redesign After Protests on Facebook and Twitter," *The Guardian*. Retrieved June 1, 2017, from www.theguardian.com/media/2010/oct/12/gap-logo-redesign.

9. James J. DiCarlo, Davide Zoccolan, and Nicole C. Rust. "How Does the Brain Solve Visual Object Recognition?" *US National Library of Medicine, National Institutes of Health*. Retrieved June 1, 2017, from www.ncbi.nlm.nih.gov/pmc/articles/PMC3306444/.

10. Josh Nofsinger. "Familiarity Bias Part I: What Is It?" *Psychology Today*. Retrieved June 1, 2017, from www.psychologytoday.com /blog/mind-my-money/200807/familiarity-bias-part-i-what-is-it.

Endnotes

11. Linda R. Weber and Allison I. Carter. *The Social Construction of Trust* (Springer US, 2003).

12. "Coca Cola vs. Pepsi: Logo Design Case Study," *Canny-Creative*. Retrieved June 1, 2017, from www.canny-creative.com/coca-cola-vs-pepsi-logo-design-case-study/.

13. LOGOBR. "Coca-Cola Brandbook—Brand Identity and Design Standards," *Issu.com*. Retrieved June 1, 2017, from issuu.com/logobr/docs/guideline_cc.

14. "KitKat," *NiceCupOfTeaandASitDown.com*. Retrieved June 1, 2017, from www.nicecupofteaandasitdown.com/biscuits/previous.php3?item=94.

15. RandomVHSRippedStuff. "KitKat (80's TV Adverts)," *YouTube*, June 18, 2010. Retrieved June 1, 2017, from youtu.be/bxJBU7Xbpzs.

16. Suzy Bashford. "Brand Health Check: Kit Kat—Has Nestle Been Wise to Repackage Kit Kat?" *Campaign*. Retrieved June 1, 2017, from www.campaignlive.co.uk/article/brand-health-check-kit-kat-nestle-wise-repackage-kit-kat-kit-kat-shed-its-old-style-foil-paper-covering-flow-wrap-tear-strip-convince-consumers-right/73773?src_site=marketingmagazine.

17. Steven E. Sexton and Alison L. Sexton. "Conspicuous Conservation: The Prius Effect and Willingness to Pay for Environmental Bona Fides," *Berkeley.edu*. Retrieved June 1, 2017, from are.berkeley.edu/fields/erep/seminar/s2011/Prius_Effect_V1.5.3.pdf.

18. K. Kim and M. K. Johnson. "Extended Self: Spontaneous Activation of Medial Prefrontal Cortex by Objects That Are 'Mine,'" *US National Library of Medicine, National Institutes of Health*. Retrieved June 1, 2017, from www.ncbi.nlm.nih.gov/pubmed/23696692.

19. Tom Bruce-Gardyne. "Hendrick's: A Brand History," *The Spirits Business*. Retrieved June 1, 2017, from www.thespiritsbusiness.com/2014/10/hendricks-a-brand-history/.

20. Hendrick's Gin website, "Our Peculiar Past" section. Retrieved June 1, 2017, from www.google.com/intl/en/about/products/.

21. Nichola Rutherford. "The Gin Crowd: Scotland's Distilleries in New Trail," *BBC*. Retrieved June 1, 2017, from www.bbc.co.uk/news/uk-scotland-35371403.

22. BMW Designworks team in discussion with Soon Yu, June 2015.

23. Mahesh Mohan. "All Things Amazon: A List of Over 21 Amazon Products & Services," *Minterest*. Retrieved June 1, 2017, from www.minterest.org/all-things-amazon/.

24. "Novelty Aids Learning," *UCL News*. Retrieved June 1, 2017, from www.ucl.ac.uk/news/news-articles/news-releases-archive /newlearning.

25. Rowan Horncastle. "Nissan Has Built a Batman-Themed Juke," *Top Gear*. Retrieved June 1, 2017, from www.topgear.com/car-news /nissan-has-built-batman-themed-juke.

26. VANS website, "History" section. Retrieved June 1, 2017, from www.vans.com/history.html#1966.

27. VANS website, "Shop" section. Retrieved June 1, 2017, from www.vans.com/shop.html.

Chapter 5

1. Natalie Robehmed. "The 'Frozen' Effect: When Disney's Movie Merchandising Is Too Much," *Forbes*. Retrieved June 1, 2017, from www.forbes.com/sites/natalierobehmed/2015/07/28/the-frozen -effect-when-disneys-movie-merchandising-is-too-much/.

2. Alison Griswold. "Disney's Frozen Sales Magic Shows No Signs of Melting Away," MoneyBox blog, *Slate*. Retrieved June 1, 2017, from www.slate.com/blogs/moneybox/2015/02/04/disney_q1 _2015_earnings_the_frozen_sales_magic_shows_no_signs_of _freezing.html.

3. "DisneyCopyright.com Marketing Guidelines," *MouseAgents.com*. Retrieved June 1, 2017, from www.mouseagents.com/wp-content /uploads/2013/12/Disney-Marketing-Guidelines.pdf.

4. Shirley Pelts. "The Walt Disney Company: Its Stock Is Magical," *Marketrealist*. Retrieved June 1, 2017, from marketrealist.com/2015 /07/disney-interactive-steps-game/.

5. Robin Lewis. "The Coming Crash of Michael Kors...Take It to the Bank," *Forbes*. Retrieved June 1, 2017, from www.forbes.com/sites /robinlewis/2014/07/17/the-coming-crash-of-michael-kors-take-it -to-the-bank/#5ac828991100.

6. Trefis Team. "Coach: Divided Against Itself?" *Forbes*. Retrieved June 1, 2017, from www.forbes.com/sites/greatspeculations/2014 /04/04/coach-divided-against-itself/#2d99a5565b89.

7. A. G. Lafley and Roger L. Martin. *Playing to Win: How Strategy Really Works* (Harvard Business Review Press, 2013).

8. Gillian Fournier. "Mere Exposure Effect," *Psych Central*. Retrieved June 1, 2017, from psychcentral.com/encyclopedia/mere-exposure -effect/.

Endnotes

9. Daniel Kahneman. *Thinking, Fast and Slow*, 1st ed. (Farrar, Straus and Giroux, 2013).

10. "Availability Heuristic," *behavioraleconomics.com*. Retrieved June 1, 2017, from www.behavioraleconomics.com/mini-encyclopedia -of-be/availability-heuristic/.

11. Collin Campbell. "'Conspicuous Conservation' and the Prius Effect," *Freakonomics*. Retrieved June 1, 2017, from freakonomics.com/2011 /04/21/conspicuous-conservation-and-the-prius-effect/.

12. Geraldine E. Willigan. "High-Performance Marketing: An Interview With Nike's Phil Knight," *Harvard Business Review*. Retrieved June 1, 2017, from hbr.org/1992/07/high-performance-marketing-an -interview-with-nikes-phil-knight.

13. "Porsche Celebrates 50 Years of the 911 With Exclusive Limited Edition Model," *Porsche*. Retrieved June 1, 2017, from press.porsche .com/news/release.php?id=789.

14. Matthew W. Hill. "A Brief History of Squier and the Origins of Fender MIJ," *Reverb*. Retrieved June 1, 2017, from reverb.com/uk /news/a-brief-history-of-squier-and-the-origins-of-fender-mij.

15. Rebecca Cope. "H&M'S Best Designer Collaborations," *Harper's Bazaar*. Retrieved June 1, 2017, from www.harpersbazaar.co.uk /fashion/fashion-news/news/g23095/hms-best-designer -collaborations/?.

16. "Beck's Art Crawl Hits Shoreditch Free Beer," *Artlyst*. Retrieved June 1, 2017, from www.artlyst.com/news/becks-art-crawl-hits -shoreditch-free-beer/.

17. Vijay Pattni. "Victoria Beckham RR Evoque arrives," *BBC Top Gear*. Retrieved June 1, 2017, from www.topgear.com/car-news/victoria -beckham-rr-evoque-arrives.

18. Carmine Gallo. "Former Apple Retail Exec Reinvents the Car Buying Experience," *Forbes*. Retrieved June 1, 2017, from www.forbes.com/sites/carminegallo/2012/01/11/former-apple -retail-exec-reinvents-the-car-buying-experience/#466ff8404d27.

19. Elon Musk. "The Tesla Approach to Distributing and Servicing Cars," *Tesla*. Retrieved June 1, 2017, from www.tesla.com/en_GB/blog /tesla-approach-distributing-and-servicing-cars.

20. Katie Fehrenbacher. "7 Reasons Why Tesla Insists on Selling Its Own Cars," *Fortune*. Retrieved June 1, 2017, from fortune.com/2016 /01/19/why-tesla-sells-directly/.

Chapter 6

1. Bloomberg News. "Inside the Converse Design Studio Where Chuck Taylors Were Reborn: 'Don't F--- With the Chuck'," *National Post*. Retrieved June 1, 2017, from nationalpost.com/life/fashion-beauty/inside-converses-design-studio-where-chuck-taylors-are-born-270000-pairs-are-sold-daily-worldwide/wcm/cc8823e4-8bdd-4eb6-90d4-b0d0da8fb647.

2. Bloomberg. "Inside the Iteration of Converse's Chuck Taylors," *Business of Fashion*. Retrieved June 1, 2017, from www.businessoffashion.com/articles/news-analysis/inside-converses-design-studio-where-chuck-taylors-are-born.

3. "Pepsi," *Logopedia* Retrieved September 1, 2017, from logos.wikia.com/wiki/Pepsi, & "Coke," *Logopedia* Retrieved September 1, 2017, from logos.wikia.com/wiki/Coca-Cola.

4. Natalie Zmuda. "What Went Into the Updated Pepsi Logo," *Ad Age*. Retrieved June 1, 2017, from adage.com/article/news/pepsi-s-logo-update/132016/.

5. Margaret Mark. *The Hero and the Outlaw: Building Extraordinary Brands Through the Power of Archetypes*, 1st ed. (McGraw-Hill Education, 2001).

6. "Porsche Torque Vectoring (PTV) and Porsche Torque Vectoring Plus (PTV Plus)," *Porsche*. Retrieved June 1, 2017, from www.porsche.com/usa/models/911/911-carrera-models/chassis/porsche-torque-vectoring-ptv-ptv-plus/.

Chapter 7

1. In-N-Out Burger website, "History" section. Retrieved June 1, 2017, from www.in-n-out.com/history.aspx.

2. Simon Martin. "What the Allure of In-N-Out Burger Can Teach Us About Becoming Legendary," *Ceros*. Retrieved June 1, 2017, from www.ceros.com/blog/in-n-out-legend/.

3. Hayley Peterson. "In-N-Out President Explains Why the Burger Chain Probably Won't Expand to the East Coast," *Business Insider*. Retrieved June 1, 2017, from uk.businessinsider.com/why-in-n-out-burger-wont-expand-east-2015-9?r=US&IR=T.

4. "Top 500 Chains of 2017: 70—In-N-Out Burger," *Restaurant Business*. Retrieved June 1, 2017, from www.restaurantbusinessonline.com/special-reports/top-500-chains/n-out-burger-2016.

5. Judy Chen, former head of marketing at Chiquita, in discussion with Soon Yu, July 2014.

6. Jay Moye. "Share a Coke: How the Groundbreaking Campaign Got Its Start 'Down Under,'" *Coca-Cola Company*. Retrieved June 1, 2017, from www.coca-colacompany.com/stories/share-a-coke-how-the-groundbreaking-campaign-got-its-start-down-under.

7. Vince Voron in a talk at FEI London, November 25, 2016.

8. Burt Helm. "How I Did It: James Dyson," *Inc.* Retrieved June 1, 2017, from www.inc.com/magazine/201203/burt-helm/how-i-did-it-james-dyson.html.

Chapter 8

1. Teressa Iezzi. "Red Bull CEO Dietrich Mateschitz on Brand as Media Company," *Fast Company*. Retrieved June 1, 2017, from www.fastcompany.com/1679907/red-bull-ceo-dietrich-mateschitz-on-brand-as-media-company.

2. Glenn Ives and Rod Thomas. "Innovation State of Play: Mining Edition 2015," *Deloitte*. Retrieved June 1, 2017, from www2.deloitte.com/content/dam/Deloitte/br/Documents/energy-resources/Innovation_State_of_Play.PDF.

3. Dave Lavinsky. "The Two Most Important Quotes in Business," *Growthink*. Retrieved June 1, 2017, from www.growthink.com/content/two-most-important-quotes-business.

4. "Jack Daniel's Whiskey," Advertisement Gallery. Retrieved September 1, 2017, from www.magazine-advertisements.com/jack-daniels-whiskey.html.

Photo Credits

Figure 11. Building with door and windows, imitating a face. Photograph by Cesare Rinaldi. Downloaded from www.flickr.com/photos /incitazioni/5475708363 in June 2017.

Figure 12. Electric Socket, imitating a face. Photograph by Fiona Zwieb in November 2017.

Figure 13. Toilet with toilet paper on top, imitating a face. Photograph by Gary Davidson. Downloaded from www.flickr.com/photos /veryscarygary/3129094726/ in June 2017.

Figure 14. Image of Dyson vacuum cleaner chamber. Photograph by razorpix from Alamy Stock Photo. Downloaded from Alamy.com in November 2017.

Figure 15. Image of Burberry signature check close up on scarf. Photograph by MARKOS DOLOPIKOS from Alamy Stock Photo. Downloaded from Alamy.com in November 2017.

Figure 16. Image of a bottle of Kikkoman soy sauce taken with a PackshotCreator photo studio by Creative Tools AB. This file is licensed under the Creative Commons Attribution 3.0 Unported license. Downloaded from commons.wikimedia.org/wiki/File:Kikkoman _soysauce.jpg in June 2017.

Figure 17. Excite! Screenshot of home page in 1996. Courtesy of WebpageFX. Image downloaded from www.webpagefx.com/blog/web -design/popular-search-engines-in-the-90s-then-and-now/ in July 2017.

Figure 18. Google Screenshot of home page. Courtesy of Google. Google and the Google logo are registered trademarks of Google Inc., used with permission. Image downloaded from www.google .com in July 2017.

Figure 19. Photo of Hendrick's Gin bottle and label. This file is licensed under the Creative Commons Attribution-Share Alike 3.0 Unported, 2.5 Generic, 2.0 Generic and 1.0 Generic license. Downloaded from Wikimedia Commons, commons.wikimedia.org/wiki/File:Hendrick %27s_Gin_1l.jpg in July 2017. Edited by Dave Birss.

Figure 20. Screenshot of Amazon 1-Click button featured on their website. Courtesy of Amazon. Image taken from www.amazon.com in July 2017.

Figure 21. Promotional photo of Nissan's limited-edition Batmobile tie-in for the film The Dark Knight Rises. Downloaded from www.autoexpress.co.uk/nissan/juke/61688/batman-inspired-nissan -juke-revealed in June 2017.

Figure 22. Vans collaborated with the acclaimed Japanese contemporary artist Takashi Murakami. Downloaded from www.highsnobiety.com/2015/06/25/vault-by-vans-takashi-murakami-collection/ in July 2017.

Figures 23–24. The Apple logo stickers placed on a journal and guitar. The Apple symbol is a registered trademark of Apple Inc. Photographed by Dave Birss in July 2017.

Figure 25. Nike Air shoe advertising campaign created by Wieden+Kennedy. Downloaded from sneakerlab.net/nike/nike-air-max-zero-white-yellow-20160122.html in July 2017.

Figure 26. Image of Tesla storefront. Courtesy of Tesla Communications. Downloaded from www.tesla.com/sites/default/files/pictures/full/retail/palo_alto.jpg?617 in July 2017.

Figure 27. The evolution of the Pepsi and Coca-Cola logos. "Pepsi," Logopedia Retrieved September 1, 2017, from logos.wikia.com/wiki/Pepsi, & "Coke," Logopedia Retrieved September 1, 2017, from logos.wikia.com/wiki/Coca-Cola.

Figure 28. Examples of Coca-Cola bottles with personalized names on them. Created by Dave Birss in July 2017.

Figures 29–31. Jack Daniel's Whiskey Taster 1961 Ad, Jack Daniel's Charcoal Mellowing Room 1973, and Jack Daniel's Woodsman 1974 Ad. Courtesy of Jack Daniel's. Downloaded from www.magazine-advertisements.com/jack-daniels-whiskey.html in August 2017.